The Human Beast

Through the Lens of Evolution

Nigel Barber

The Human Beast

Printed in the United States of America

Publisher, Trudy Callaghan, Portland Maine

Cover Design: Trudy Callaghan

ISBN: 978-0-9855691-2-9

Dedicated to my granddaughter

Beatrix Barber-Callaghan

Other Works by Nigel Barber

Why Parents Matter

The Myth of Culture

The Science of Romance

Kindness in a Cruel World

Evolution in the Here and Now

Why Atheism Will Replace Religion

Parenting: Roles, Styles & Outcomes

Encyclopedia of Ethics in Science & Technology

Contents

Contents (continued)

"Nothing in Biology Makes Sense Except in the Light of Evolution"

- Theodosius Dobzhansky, evolutionary biologist, 1973

Why I Wrote This Book

These short pieces were published as blog posts on Psychology Today. The topics are wide-ranging and this miscellany was selected by Trudy Callaghan, the editor. Although varied, all have a common genesis and method of development.

I began with what seemed like an interesting question that piqued my curiosity but seemed to lack any obvious, or widely-accepted answer. This is a risky endeavor and it is liable to fail when the answers are incomplete, unsatisfactory, or too speculative. As often as not, I was pleasantly surprised by the outcome – if not, no one needed to know!

My approach uses evolutionary thinking that focuses on humans as an evolved species. While this perspective is often portrayed as limiting, it is quite the opposite. Evolutionary theory is like an all-purpose telescope that can focus on any point in the firmament. Whether the topic is irrational fears, artistic ability, or the impact of the weather on our moods, we are treading territory already visited by animal behaviorists and evolutionary theorists. This perspective offers many surprising, and revealing insights into what it means to be human.

Nigel Barber, Ph.D.

Is Sports a Religion?

Have You Ever Noticed that Sport is Full of Rituals

Psychologists are closing in on the conclusion that sport has many of the same effects on spectators as religion does. Here is Daniel Wann[1], a leading sport psychologist at Murray State University, and his co-authors:

> *The similarities between sport fandom and organized religion are striking. Consider the vocabulary associated with both: faith, devotion, worship, ritual, dedication, sacrifice, commitment, spirit, prayer, suffering, festival, and celebration. p. 198*

It may seem odd to equate religion with sport entertainment, but it must be understood that prior to mass communications, religious ceremonies were a source of entertainment for ordinary people who rarely attended a theater or traveled to a sporting event. Sports and religion may get categorized separately but their intersection is difficult to miss.

As Wann and collaborators note, various scholars discuss sport in terms of "natural religion," "humanistic religion," and "primitive polytheism" pointing out that "spectators worship other human beings, their achievements, and the groups to which they belong." And that sports stadia and arenas resemble "cathedrals where followers gather to worship their heroes and pray for their successes."[1]

If ritual may be entertaining, then entertainment, as experienced in a sports stadium, may be ritualistic. Fans wear the team colors and carry its flags, icons, and mascots. Then there is repetitive chanting of team encouragement, hand-clapping, booing the other team, doing the wave, and so forth. The singing of an anthem at a sporting event likely has similar psychological effects as the singing of a hymn in church.

Given that sports entertainment has obvious similarities to religious rituals, it is reasonable to ask whether the connection between fans and their preferred sport has psychological effects that are comparable to religious experiences - effects that account for religion as a worldwide human adaptation.

Sports as a substitute for religion

As a group, sports fans are fairly religious, according to research. It is also curious that as religious attendance rates have dropped off in recent decades, interest in sport spectatorship has soared. Moreover, research has debunked several stereotypes about sports fans that seem incompatible with religiosity. Fans are not lazy, nor are they particularly prone to violence. Male fans do not have bad marriages.

Some scholars believe that fans are highly committed to their favored stars and teams in a way that gives focus and meaning to their daily lives. In addition, sports spectatorship is a transformative experience through which fans escape their humdrum lives, just as religious experiences help the faithful to transcend their everyday existence.

From that perspective, the face painting, hair tinting, and distinctive costumes are thought to satisfy specific religious goals including identification with the team, escape from everyday limitations and disappointments, and establishing a community of fans.

So far, the transformative aspects of fandom are quite close to those associated with religion. Lest the fans become too smug, here is a socialist critique:

Shaped by the needs of capitalist systems, spectator sports serve vested interests as a type of "cultural anesthesia," a form of "spiritual masturbation," or "opiate" that distracts, diverts, and deflects attention from the pressing social problems and issues of the day[1].

2

Of course, Karl Marx famously declared that religion is the opium of the people. Not all religions numb people to their social and moral responsibilities, however. One thinks of liberation theology in Latin America, for instance. No one ever claimed that sports had comparable redeeming qualities. According to one critic (Harris, 1981)[2], "it has turned into a passion, a mania, a drug far more potent and widespread than any mere chemical substance." It is the new opium of the people.

Sources

1. *Wann, D. L., Melznick, M. J., Russell, G. W., & Pease, D. G. (2001). Sport fans: The psychology and social impact of spectators .New York: Routledge.*
2. *Harris, S. J. (1981, November 3). Sport is new opium of the people. Democrat and Chronicle, p. 3B.*

Belief in Brands is More Emotional than Rational
Why brands shape feelings and simplify purchase decisions

W hen customers buy luxury brands, they are investing as much in the reputation as in the product. This phenomenon is built on ancient social conformity and consumers pay through the nose for a benefit that is often more social and cognitive than material.

Brands as Belief Conformity

A brand is essentially a self-fulfilling prophecy. Consumers learn that a certain brand is the most popular one. They assume this must reflect superior quality. This belief reinforces repeat purchases as the consumer develops brand loyalty.

Brand loyalty is partly the product of a Darwinian struggle among rivals where products having design weaknesses get selected out ensuring that surviving brands are of minimally good quality.

Brand loyalty is often compared to personal loyalty and marketers even speak of brands having "personality." A brand of footwear might be associated with rugged individuality and with healthy living, for example. When a brand personality is consistent with a person's own personality, this strengthens brand loyalty

There is much about brand loyalty that is highly irrational but it is the kind of irrationality that defines humans as a social species. One of the more obvious features of brand fidelity is that it is based upon social conformity. Social conformity can help us to thrive in many different settings. We are all more likely to survive if we agree on rules like driving on the same side of the road.

The downside of conformity is that poor decisions are often socially propagated which may be why entertainment that is mediocre, formulaic, and incites visceral thrills makes more money than alternatives that call for deep reflection.

The Wisdom of the Crowd

Although group decisions often favor mediocrity, they have the advantage of being mostly more right than wrong. From that perspective, buying a respected brand is one method of handling the indecision of selecting a product.

If a self-driving car crashes into a tree and bursts into flames, the manufacturer's brand loyalty and brand value suffer. Similarly, if exercise equipment injures a child, this is very bad for business. If products generate such problematic outcomes, they are unlikely to be, or to remain, top brands. So, going with dominant brands can be interpreted as a form of risk management.

While it makes sense to pay more for brand names if they guarantee some level of quality, people pay far more for top brands than this cognitive decision-making aspect would appear to merit, particularly in the Internet age when most information is free.

The Role of Price and Advertising

Why does a luxury brand cost so much more than a "Brand Y" alternative? There may well be an added element of quality control, but this is not always true.

Indeed, for some products, such as fashionable shoes, there is generally some distinction on the basis of styling but very little difference in manufacturing quality. In many cases, luxury brands and cheaper shoes are made from the same materials using the same processes at the same factory, possibly in China.

The difference between name-brand products and their lesser known competitors often reduces to the brand itself. For its part, the brand is largely a product of marketing. Companies spend a lot of money on advertising and promotions to maintain the strength of the brand and the loyalty of customers.

In essence, customers value a product largely because companies spend a lot of money convincing us how good it is. This is highly irrational from an economic perspective but makes more sense when one considers that the consumer is not purely interested in buying quality but also purchases status and self-esteem by advertising their capacity to buy "the best."

The Fearless Consumer

Consumerism is a philosophy of life that springs into existence as societies become more economically developed, and as discretionary income rises. In this social environment, most purchases are disconnected from basic biological needs and serve instead to satisfy psychological needs.

These include a desire to communicate high social status and success as well as being fashionable and politically astute by avoiding products that have unpleasant connotations, such as furs from endangered species, and gas-guzzling cars.

Ironically, the entire thrust of consumer societies is towards over-consumption of goods and services that raise carbon emissions and endanger the planet. This contradiction spurs many producers of name-brand goods to emphasize their climate awareness and to publicize their efforts to establish circular economies and generally reduce their carbon emissions. To do less would be to bring their operations into disrepute and to injure brand loyalty and brand value.

Source
https://www.ingentaconnect.com/content/sbp/sbp/2015/00000043/00000009/art00002;jsessionid=3fqlql9oatb9c.x-ic-live-01 Consistency of brand personality with own personality strengthens brand loyalty

Can Capitalism be Fair?

The two major challenges of capitalism are wealth inequality and climate change

The key problem with capitalism is that it involves a conflict between workers and owners. We see this issue today in the increasing luxury of owners and the bitter economic struggles of employees. The last time inequality was so extreme, we got the political turmoil preceding World War II. From early in its history, capitalism manifested a capacity to undermine the power of governments and citizens alike.

The Joint Stock Company

In England, the turn of the eighteenth century saw the inauguration of the British East India Company by royal charter. The purpose of the company was to exploit resources from Asia, and it was quickly followed by the Dutch East India Company.

The British East India Company was more powerful than most governments of its day. It set up trading posts around Indonesia and China and hired a large army to defend them. This company functioned as a vehicle for colonial domination of distant populations (including India) and offers a frightening illustration of what happens when capitalism operates outside of government authority. This is clearly relevant to our own time where large global companies no longer need armies to exert their will around the globe.

The British East India Company returned a fat dividend that enriched investors for generations. Then it stopped being profitable due to the spiraling costs of empire and went out of business to be replaced by direct control from the London government.

This early company suffered from most of the problems of modern capitalism, including the raping of global resources and the obscene accumulation of wealth in the hands of the few.

Capitalism and Inequality

The wealthiest one percent of the U.S. population own about half of the financial wealth. So, the financial markets are the primary source of wealth inequity. Such extreme inequality poses threats to democracy for several different reasons.

One is that wealthy citizens can have an outsize influence in elections by making large contributions to election campaigns that often purchase access to elected officials. Another is that instead of being a merit-based society where anyone can succeed by virtue of their own efforts, the U.S. is becoming a society of hereditary wealth where children of the affluent have no need to work and may contribute little to their country or community. (Of course, the same is true of some nominally socialist countries like China that have adopted a capitalist business system).

The current level of inequality is equivalent to that of 1929 and capitalist societies that are this unequal generally prove unstable.

There are two key sources of instability. The first is political and one gets revolutions on the right and the left such as those occurring in the decade of the 1930s. There are currently many signs of political instability in countries around the globe that are fueled by economic unrest, giving rise to populist revolts.

Inequality is itself a cause of economic failure because the relative poverty of large segments of the society crimps demand for goods and services. The most under emphasized source of instability in capitalist societies may be damage to the environment and ecosystems. This has recently been emphasized by climate change activists.

Capitalism and Ecological Collapse

Capitalist societies thrive by increasing economic production, raising wages, and increasing the standard of living. This trend has continued since the Industrial Revolution so that workers have ever-increasing disposable income and generally improving health as measured by an approximate doubling of life expectancy[1].

Most of the increases in disposable income show up as rising consumption of goods and services. Increased consumption boosts the use of energy and other resources.

For example, the furniture placed in larger modern houses is produced from hardwoods taken out of old-growth forests around the globe that compromises the planet's capacity to store carbon, substantially increasing green house gases and climate change.

While the burden of climate change is placed on everyone to some extent, the biggest losers often receive few of the gains of the planetary pillage. For example, many poor low-lying countries in the Pacific stand to lose substantial territory to rising sea level.

Another problem is that the advantages of capitalism are front-loaded. They are enjoyed today whereas the biggest costs are borne by future generations whether as a direct result of climate change, or through loss of biodiversity and collapse of economically important ecosystems, such as the world's fish stocks.

Generational Competition

This issue recently came to the fore in the youth environmentalism movement with high school students demanding that governments take responsibility for correcting climate change instead of pushing the problem off to future generations who must cope with a substantially hotter world.

The character of this competition can be simply stated. Current generations are over- consuming planetary resources causing inevitable future damage to climate and ecosystems.

To the extent that this damage is permanent and irreversible, the excessive consumption of current generations on impractically large homes, heavy vehicles, luxury goods, luxury travel, inessential goods, consumer products, and reckless consumption of energy, is a price exacted on innocent future generations.

Small wonder that they would interpret the narrative of endless economic growth as a fairy tale.

Sources

1. Floud, R., Fogel, R. W., Harris, B., & Hong, S. C. (2011). *The changing body: Health, nutrition, and human development in the Western world since 1700.* Cambridge, England: NBER/Cambridge University Press.

Why Atheism Will Replace Religion
The Evidence

A theists are heavily concentrated in economically developed countries, particularly the social democracies of Europe. In underdeveloped countries, there are virtually no atheists. Atheism is a peculiarly modern phenomenon. Why do modern conditions produce atheism? In a new study, I provide compelling evidence that atheism increases along with the quality of life[1].

First, as to the distribution of atheism in the world, a clear pattern can be discerned. In sub-Saharan Africa there is almost no atheism[2]. Belief in God declines in more developed countries and atheism is concentrated in Europe in countries such as Sweden (64% nonbelievers), Denmark (48%), France (44%) and Germany (42%). In contrast, the incidence of atheism in most sub-Saharan countries is below 1%.

The question of why economically developed countries turn to atheism has been batted around by anthropologists for about eighty years. Anthropologist James Fraser proposed that scientific prediction and control of nature supplants religion as a means of controlling uncertainty in our lives. This hunch is supported by data showing that the more educated countries have higher levels of non belief and there are strong correlations between atheism and intelligence.

Atheists are more likely to be college-educated people who live in cities that are highly concentrated in the social democracies of Europe. Atheism thus blossoms amid affluence where most people feel economically secure. But why?

The Case for Affluence

It seems that people turn to religion as a salve for the difficulties and uncertainties of their lives. In social democracies, there is less fear and uncertainty about the future because social welfare programs provide a safety net, and better health care means that fewer people can expect to

11

die young. People who are less vulnerable to the hostile forces of nature feel more in control of their lives and less in need of religion. Hence my finding that belief in God is higher in countries with a heavy load of infectious diseases.

In my 2011 study of 137 countries[1], I also found that atheism increases for countries with a well-developed welfare state (as indexed by high taxation rates). Moreover, countries with a more equal distribution of income had more atheists. My study improved on earlier research by taking account of whether a country is mostly Moslem (where atheism is criminalized) or formerly Communist (where religion was suppressed) and accounted for three-quarters of country differences in atheism.

In addition to being the opium of the people (as Karl Marx contemptuously phrased it), religion may also promote fertility, particularly by promoting marriage[3]. Large families are preferred in agricultural countries as a source of free labor. In developed countries, by contrast, women have exceptionally small families. I found that atheism was lower in countries where a lot of people worked on the land.

Even the psychological functions of religion face stiff competition today. In modern societies, when people experience psychological difficulties they turn to their doctor, psychologist, or psychiatrist. They want a scientific fix and prefer the real chemical medicines dished out by physicians to the metaphorical opiates offered by religion. No wonder that atheism increases along with third-level educational enrollment[1].

The reasons that churches lose ground in developed countries can be summarized in market terms. First, with better science, and with government safety nets, and smaller families, there is less fear and uncertainty in people's daily lives and hence less of a market for religion. At the same time, many alternative products are being offered, such as psychotropic medicines and electronic entertainment that have fewer strings attached and that do not require slavish conformity to unscientific beliefs.

Sources

1. Barber, N. (2011). *A cross-national test of the uncertainty hypothesis of religious belief.* Cross-Cultural Research, *45, 318-333.*

2. Zuckerman, P. (2007). *Atheism: Contemporary numbers and patterns. In M. Martin (ed.), The Cambridge companion to atheism. Cambridge: Cambridge University Press. This book is not held by any U.S. Library.*

3. Sanderson, S. K. (2008). *Adaptation, evolution, and religion. Religion, 38, 141-156.*

Dog Ownership and Health

Dogs improve health behavior more than they change health as such

It was startling to read recent exaggerated claims that owning a dog could increase life expectancy by about ten times as much as being married. The findings provide a cautionary tale in how scientific results are interpreted.

The News May be too Good

The results of a recent study published by the Mayo Clinic suggest that some people live much longer if they are dog owners. The study population were Swedish heart attack and stroke patients.

This means that the findings tell us little, or nothing, about the general population. It is a truism of statistical inference that the results for a specific population cannot be extended to different populations.

Even so, the numbers are startling. The study reports that heart attack and stroke patients live more than 20 percent longer if that are dog owners compared to non dog owners.

To put this in context, we are dealing with a boost to health that is an order of magnitude greater than the lifelong benefit of being married. As such, it seems too good to be true. While the study seems to have been rigorously conducted, it is possible that something went wrong.

Once again, basic statistical inference could be in play. One anomaly of the study was extremely low dog ownership rates of study participants. At less than 5 percent, the dog ownership rates were strikingly low in a country (Sweden) where 24 percent of the population owns dogs according to recent surveys.

The criteria of dog ownership may have been too restrictive, introducing a distortion in the study if the population of dog ownership among heart and stroke patients was under represented and non representative. Whatever about this clinical population, there are good reasons for believing that dog ownership can confer health benefits.

The Nitty Gritty of How Dogs Affect Health

Owning any pet may well confer psychological benefits and improve somatic health. Yet, the focus on dogs is not difficult to explain.

To begin with, dogs are mostly very generous with their affection. They provide positive interactions that are reliable and have little to do with how the owner feels, or acts. Unconditional affection is the gold standard by which intimate emotional relations among people are judged.

Such relationships protect us against loneliness and depression so that dog ownership can produce beneficial psychological outcomes. Another key reason that dog ownership contributes to health is that it makes people more physically active. This helps explain why health researchers have zeroed in on canine companions rather than other common pets, such as cats.

Most cats do not need to be walked and most also show limited affection. The affection expressed by cats is also less predictable, evidently depending upon how the individual animal feels at a specific time. These factors may be why cat ownership attracts less attention by health researchers. What of the evidence for health effects from dogs?

Health Effects in a Random Sample of Czechs

Given that dog owners walk more, and spend more time outdoors in pleasant natural scenery, one might expect them to have better cardiovascular health. That is indeed the case. That conclusion is widely-known and was confirmed in a high quality study conducted in the city of Brno in the Czech Republic (Czechland).

This study improved over others by randomly selecting subjects and conducting lab measurements related to cardiovascular health. Using this superior methodology, researchers found substantial health advantages among dog owners. Yet, the benefits of dog ownership on health disappeared when controlling for age, activity level, and socioeconomic status of the participants.

Reading between the lines, it appears that much of the health advantages of dog ownership overlap with the benefits of exercise in improving cardiovascular health. Owning a dog motivates owners to get more exercise so that they can expect to be healthier. While owning a dog generally contributes to health, it can also have downsides.

How Dog Ownership Can Undermine Health

Dog ownership can be a source of health problems. To begin with, older people are more vulnerable to tripping and falling particularly when walking an enthusiastic large canine. Dogs have short lives compared to people so that their loss is a cause of recurrent bereavement that can trigger depression in vulnerable individuals.

Domestic dogs suffer a great many illnesses that are almost as expensive to treat as human health problems. Even when healthy, a dog is a drain on finances that may be a significant burden for low-income households. Financial strain is a major source of stress in people's lives and can undermine health.

The responsibility of caring for a pet can be socially isolating if owners feel obliged to stay at home to care for their animal. Social isolation is also a threat to health.

Despite such problems, the recent findings suggest that dog ownership can boost health substantially. This benefit probably reflects higher levels of physical activity.

Sources

https://mcpiqojournal.org/article/S2542-4548(19)30088-8/fulltext Benefits of dog ownership on health disappeared when controlling for age, activity level, socioeconomic status. This was a random sample of residents from Brno Czech Republic

https://www.ahajournals.org/doi/10.1161/CIRCOUTCOMES.118.005342 Heart attack and stroke victims survived much longer if they owned a dog in Sweden where less than 5 percent of patients owned a dog

https://www.statista.com/statistics/917563/pet-ownership-in-sweden-by-pet-type/ 24 percent of Swedes own dogs according to Statista (2018)

Real Reasons for Sex Before Marriage
The practical reasons for premarital sex trump religion and morality

More people today are sexually active before marriage than ever before. The true reasons are practical and have little to do with changing belief systems.

In America a century ago, only a small minority of women were sexually active before marriage (about 11 percent)[1], compared to a large majority today. The same pattern is observed in other developed countries.

Why are women in developed countries more sexually active before marriage? A long list of practical explanations cut across all belief systems:

Contraception
With the wider use of effective contraceptives, young women do not fear unwanted pregnancy so much as earlier generations did.

The key event here was the widespread adoption of the contraceptive pill in the early 1970s. Because this was highly effective and female-controlled, it took away most of the anxiety about unwanted pregnancy.

Reduced Parental Supervision
Teens are less supervised after school if both parents work full-time and may take advantage of this opportunity for sexual activity. Increased enrollment in higher education means that a lot of young women live apart from their families in an environment that encourages sexual expression. This is in marked contrast to sexually restrictive societies, where young, single women are heavily chaperoned by relatives.

Earlier Sexual Maturation of Women and Later Age of Marriage
In the 1860s, women did not mature reproductively until the age of 16 years, compared to 11-12 years today[2]. First marriages are later

today also, with European women postponing matrimony until the age of about 29. So, there is a very long interval of about 10-20 years between puberty and marriage during which complete sexual abstinence is unlikely.

More Women in the Workforce

As more women enter paid employment and careers, they spend more time preparing for the workforce through third-level education. The number of single, never-married young women is on the rise. Most of these women are sexually active.

More Gender Equality in Jobs

Women used to be far more economically dependent on fathers and husbands. With greater economic independence and more female-headed households, women are freer to control their sex lives, as feminist writers like Helen Gurley Brown[3] pointed out. This means more premarital sex and increased single parenthood.

Women Are More Competitive and Sensation-Seeking

Contemporary women are more competitive in a number of arenas, from sports to education, politics, and careers. Competitiveness is associated with a hormone profile of high sex drive in both sexes[4]. Women's risk profile is converging with that of men, as illustrated by rises in problem drinking and dangerous driving. They are also less risk-averse in sexual matters, increasing premarital sexuality.

Declining Marriage

About a fifth of American women never marry[5]. Of those who do, the chances of remaining married to the same person for life are low. The time spent in marriages is decreased by divorce, even if most divorcees remarry. In the U.S., close to half of first marriages end in divorce, and the typical duration of a first marriage is only seven years. Between non-marriage, late marriage, and frequent divorces, larger numbers of women live as singles than ever before, boosting premarital sex.

The Mate Market

A large number of sexually active, single women means that men do not need to marry to enjoy an active sex life. If a man may sleep with various attractive women without any long-term commitment, he is less likely to propose marriage to any of them. So, romantic relationships are negotiated on the basis of what typical men want, which is sex early in a relationship with little in the way of a permanent commitment such as marriage. (The irony is that with "sexual liberation," women lost power in relationships at the same time that they gained power in the economy.)

In a published paper, I tested out some of these ideas in a comparison of 40 countries[6]. My analysis looked at acceptance of premarital sex (that is almost perfectly correlated with self-reported sexual behavior). I found that premarital sex increases in more developed countries that have higher-paid labor force participation by women. Premarital sex increases in countries having weak marriages (i. e., low marriage rates, and high divorce rates). Countries where more children are born outside of marriage are more accepting of premarital sexuality, but very religious countries strongly reject it. Some, like Pakistan, have severe penalties for premarital sex, up to and including so-called honor killings. In such cases, the impact of religion on sexual behavior is likely more practical than doctrinal. If so, it fits the pattern of sexual behavior adapting to local costs and benefits.

Sources

1. *Caplow, T., Hicks, L., & Wattenberg, B.J. (2001). The first measured century: An illustrated guide to trends in America, 1900-2000. La Vergne TX: AEI Press.*
2. *Daly, M.,Wilson, M. (1983). Sex, evolution, and behavior. Belmont, CA*
3. *Gurley Brown, H. (1962). Sex and the single girl. New York: Bernard Geis.*
4. *Cashdan, E. (2008). Waist-to-hip ratios across cultures: Trade-offs between androgen- and estrogen-dependent traits. Current Anthropology, 49, 1099-1107.*
5. *Klinenberg, E. (2012). Going solo: The extraordinary rise and surprising appeal of living alone. New York: Penguin.*
6. *Barber, N. (2017 b). Cross-national variation in attitudes to premarital sex: Economic development, disease risk, and marriage strength. Cross-Cultural Research, 1-15.*

Is Psychology Responsible for the Unabomber?
" Manhunt" documentary partly blames unethical research

Psychology professors like to expose unethical research. Yet, Harvard professor Henry Murray - a leading motivation researcher with a cavalier approach to ethics - passed under the radar. He even received two prestigious awards from the American Psychological Association (APA).

The Ethics Atrocity

U.S. intelligence services wanted to devise a procedure for "breaking" cold war Soviet spies so that they would be completely useless to their handlers. They hired Henry Murray to come up with an effective method of doing this.

Murray deceived Harvard psychology students into thinking that their participation would make a huge contribution to scholarship. He used personality tests to recruit emotionally vulnerable students and proceeded to undermine their egos. He did this by subjecting them to stressful interviews delving into their past. The recorded interviews were subsequently replayed to the subjects amid humiliating criticism by a trained confederate posing as a subject.

One of Murray's subjects was math major Theodore Kaczynski, a.k.a the Unabomber.

Harvard and the Unabomber

This connection was described in a 2003 book, *Harvard and the Unabomber*[1,] which was a source for the 2004 Netflix series, *Manhunt*, that repeats controversial claims about the impact of Murray's unethical research.

To summarize, the argument is made that the brilliant, but troubled, math major was turned from a sweet and precocious kid into an angry terrorist by his research experiences. Kaczynski was certainly extremely angry about his three years of deliberate psychological abuse in Murray's lab. This anger motivated his willingness to blow up complete strangers in a quest for revenge upon universities and symbols of the Industrial Revolution (that Kaczinski blamed for turning people into mindless rule-following automatons who obeyed authority figures like bosses and researchers).

By tolerating abusive research procedures, did Harvard University create a terrorist?

Can We Blame Harvard for the Unabomber?

Blaming Harvard for the Unabomber is going too far for several different reasons. The first is that we know far too little about the prediction of individual behavior to make any such claim. Moreover, he showed early signs of social difficulties that were subsequently attributed to paranoid schizophrenia.

Another issue is that Kaczinski complained bitterly about his own lifelong betrayal by family and acquaintances that fit the mold of his experiences with Murray. Whether he truly was a victim in this sense is debatable, but his life certainly offers no scarcity of supportive material, that would have been exaggerated by his paranoid tendencies.

So, if he had never met Henry Murray, his life might conceivably have played out similarly. His much-discussed alienation from modern life was not peculiar to the Unabomber but was common amongst other Harvard students and intellectuals of his day[1].

In his own mind, the Unabomber was a victim of betrayal by virtually everyone he knew - childhood friends, a young woman who rejected his romantic advances, his parents who sent him to Harvard at the age of 16 before he was emotionally mature enough to prosper there, the other Harvard students, and, finally, his brother who had supported him throughout his life. So, although the Harvard experience as a research subject probably did not create the Unabomber, it was likely a factor.

Ethical Principles

Blaming Harvard for the Unabomber may be a stretch. Yet, the fact that this case has been made by serious writers highlights the importance of ethics in research.

As an assault on the psychological integrity of the individual, Murray's research was clearly beyond the pale and ethically indistinguishable from the gruesome biomedical experiments perpetrated against concentration camp inmates by Josef Mengele and others[2].

Such atrocities precipitated the Nuremberg Code that is the guiding light for ethical research in all disciplines, including psychology.

Amongst its provisions is the notion of informed consent which Murray followed in a quite superficial manner. His participants did sign consent forms but were not informed of the true purpose of the research, much less of its goal of shattering the ego to the point that subjects would be unable to conduct a normal work life (whether as spies or as anything else). Consent may have been obtained but it was not fully informed.

Although participants were free to leave whenever they wished, Murray rendered that unlikely through a deliberate grooming process where they were led to believe that their participation had an important scientific objective. Participants were paid and Kaczinski likely stayed on because he needed the money. Murray initially lied about the length of the study stating that the commitment was for a single year.

Under current ethical rules, the use of deception requires full debriefing so that the subjects are disabused of any misleading information used by the researcher. Apparently, this was not done, and could not have been done in principle as the research was classified seems. It seems the lead researcher had a security clearance.

Of course, the central ethical lapse was ignoring the principle that no permanent harm should be done to participants (the right of non-maleficence). If such harm is done, an Institutional Review Board would have to establish that the benefits of the research greatly outweigh the harm to the subjects. In this case, given that the research was of zero value to science, this bar would not have been met.

Murray's research could not be conducted today. That begs the question of why the APA has not stripped Murray posthumously of his awards. These now seem as absurd as Yasser Arafat getting the Nobel Peace Prize.

Sources
1. Chase, A. (2003) Harvard and the Unabomber. New York: W. W. Norton.
2. Barber, N. (2002). Encyclopedia of ethics in science and technology. New York: Facts on File.

How Men Attract Women

Methods extend from animal behavior to clothes shopping

How men attract women is a much more difficult question than vice versa because women are both more selective and also more individualistic in what they want.

How Many Criteria Count?

Men are evaluated as aesthetically less pleasing than women in research using male and female raters of photographic images. This suggests that physical appeal was not selected so strongly for men as for women in our evolutionary past[1].

Even if sexy appearance is less important for men, it still matters a lot. In many cases, women react more strongly to negative traits than positive ones. Some are turned off by baldness, some reject short men, and others are most put off by a protruding stomach[2].

Reading between the lines, most women are attracted to strong, healthy, physically fit specimens who project confidence and are more likely to succeed in surviving, reproducing, and prospering in any society.

Physical attractiveness is more critical early in a relationship - presumably because it colors first impressions so much - and women who are interested in a short-term relationship are likely to have their fling with someone who is above average in physical appeal.

Once a man passes the first screen of physical attractiveness, a woman is likely to pay more attention to personality characteristics, intelligence, and general suitability for a relationship.

Idiosyncrasy Rules

The key traits of kindness and generosity are important to men as well as women in their search for a mate. This makes sense considering that any relationship is a partnership that functions best if each takes account of the interests of the other.

Even so, many skilled womanizers are far from being nice people. Instead, they are successful salespeople who succeed mainly by telling their mark what she wants to hear. Needless to say, this strategy has a limited shelf life and torrid romances often precede equally heated split-ups.

Whereas most men may fall for a stereotypically attractive woman, women are more idiosyncratic in their tastes, possibly because they take so many criteria into account. Some are wowed by bookish college professor types. Others are drawn to the physical charisma of a football player. Many gravitate to successful artists and musicians.

Most women avoid unemployed drunks and seek out partners who have at least good prospects for gainful employment. They use an economic filter for selecting a mate - even today in online dating[3]. This makes sense because getting united with success is a better recipe for the future than getting hitched to failure.

Even given the economic pickiness of most women, some set much more stringent criteria than others and it helps if they have high mate value themselves. For instance, Marilyn Monroe targeted men at the top of the tree from baseball legend Joe DiMaggio to President John F. Kennedy[2].

In general, women are drawn to people who are from similar backgrounds in terms of ethnicity, politics, and religion; and researchers find that married couples are remarkably similar in about every trait one can measure, from intelligence to relative height for gender[2].

What's a Guy to Do?

When women want to attract a mate, they lavish a lot of attention on their appearance, buying new clothes and working hard on grooming and makeup. In contrast, men expend a lot less effort on their appearance, although any effort to appear healthy and physically fit is well spent.

Clothing is important also, and although men spend considerably less than women on apparel, they spend substantially more if they are actively dating as opposed to being married.

One reason that dress is important for men is that it projects social status; women who reject a man in a Burger King uniform might be willing to date someone in medical scrubs. The quality of non-work clothing also conveys an impression about where a person shops and their disposable income.

Women are attracted to other signs of high social status, from elite diction to driving an expensive car, or dining at a posh restaurant[3]. Perhaps, for this reason, many still prefer if the man pays the expenses of a date. In our society, this is a sign of having disposable income and being generous enough to dispose of it.

Even in societies that did not use money, men were expected to bring gifts when they courted. This practice bears analogy with the nuptial gifts of birds and insects where the female is lulled into a sense of security by getting a morsel to eat.

The act of generosity is invariably by the male. This fits in with Bateman's Principle according to which males are more eager to mate and therefore must do something to win over the female. So men can bring gifts, including gifts of food.

Sources

1. Barber, N. (1995). *The evolutionary psychology of physical attractiveness. Ethology and Sociobiology, 16, 395-424.*
2. Barber, N. (2002). *The science of romance.* Buffalo, NY: Prometheus.
3. Hitsch, G., Hortacsu, A., & Ariely, D. (2010). *What makes you click?—Mate preferences in online dating. Quantitative Marketing and Economics, 8, 393-427.*

What Behaviors Do We Inherit Via Genes?
There is no gene for cleaning out the refrigerator

A pervasive assumption in evolutionary psychology is that how we act is affected by the genes we carry. Is there good concrete evidence of this? Are our outcomes predetermined by our biology? The most intriguing findings on this issue came from twin studies.

Evidence that Genes Affect Human Behavior

The study of identical twins reared apart is a natural experiment where two individuals with exactly the same genes grow up in different environments. If they turn out to be similar, then the similarity can be attributed to genotype.

Behavior geneticists concluded that genetics plays a big role in personality, accounting for about half of the differences in personality test results and even more of the differences in IQ scores.

Apart from these scientific findings, researchers were impressed by many obvious similarities between twins when they were reunited for the first time after being separated from birth. Many of the pairs dressed similarly or had the same haircut, or glasses. They described remarkable similarities in hobbies and interests. One pair reported that they were the only ones in their neighborhood to construct a circular bench around a tree in their backyard.

Striking as such stories are, they remain mere anecdotes and have no scientific value. The main problem is that there is confirmation bias. If a pair of twins is wearing the same baseball hat, we tend to interpret this as a wonderful example of genetic control over the minutiae of behavior. If a pair shows up wearing different hats, however, we ignore that difference but instead register some similarities such as both twins wearing a black shirt.

Identical twins separated at birth have some striking differences. If one twin is schizophrenic, there is no more than a coin-toss chance that the other is diagnosed with the same mental disorder. This is striking given that schizophrenia is believed to have a basis in brain biology. (The same is true of political affiliation).

We must also recognize that identical twins are a special case whose relevance to the behavior of ordinary people is disputable. The problem is that many characteristics are affected by multiple genes. If there are six genes involved, identical twins will be the same because they have all six genes. Yet, taken separately, each of those genes might not have a detectable effect on the trait of interest if studied in the general population.

This wrinkle (known as epistasis) may help explain why it is so difficult to establish a biochemical chain of causation between specific genes and complex human behaviors, although researchers have made heroic efforts to account for various traits, such as sensation seeking as a function of dopamine receptors and have investigated various candidate genes to account for criminal violence.

Biochemistry and Behavior

Establishing that some behavioral traits are heritable is not the end of the scientific mission but really just the beginning. We need to know not just that genes affect behavior but also have to establish which genes are involved and how they affect the biochemistry of brain cells in ways that influence behavior.

One of the first of such projects involved work on receptors for dopamine that are implicated in sensation seeking.

This research proved successful. Yet, the success was qualified because variation in the dopamine receptor explained only a tiny fraction of individual differences in the sensation-seeking trait.

Another study looked at the so-called "warrior genes" that were over-represented among violent criminals. Criminal defense attorneys were excited by this finding because it offered a new defense strategy for violent offenders, namely that they were not fully responsible for their actions because their genes made them do it.

That genetic defense has been a flop, however. Warrior genes affect violent behavior only in the small category of individuals who grow up in extremely abusive homes. Children who are raised by loving parents are very unlikely to engage in orgies of uncontrolled aggression.

So there is a striking contradiction between the seeming clarity of the early research via twin and adoption studies, that established clear and substantial effects of genetic inheritance on personality and behavior, and subsequent efforts to work out how these influences play out.

Adaptation Without Genes

Although it is hard to deny genetic influences on human behavior, anyone who tries to explain what a person does in terms of simple biochemical differences is likely to be disappointed. Personality psychologists recognize that gene effects are difficult to separate from environmental influences. Children growing up in the same home experience that environment very differently because they have distinct temperaments, are treated differently by parents and siblings, and pursue different interests with different companions.

For example, a child with a greater sense of curiosity is going to cultivate varied interests and activities that feed the thirst for knowledge, whereas less curious siblings extract far less intellectual stimulation from their home environment. Such differences between siblings in what they get out of the environment are about as important as genes in determining personality and intelligence[1].

So, there is little doubt that how we act is affected by genes in fairly generalized ways. Some individuals are born with a propensity to be outgoing, to be happy, emotionally reactive, sociable, creative, or intelligent. Yet, we do not have a good understanding of any of the relevant biochemical mechanisms.

Moreover, there is no satisfactory explanation of the underlying biochemical mechanisms in most cases. There is an important distinction between personality predispositions and actual behavior. Personality may be genetically heritable to some degree, but human behavior never is.

Honeybees have a complex sequence of hygienic behavior that consists of digging out infected larvae and chucking them out of the hive - a sequence that is understood in terms of Mendelian genetics with one gene for uncapping and another for removing the dead larvae[2]. As far as humans are concerned, we may or may not have strong hygienic tendencies, but there is no gene for cleaning out the refrigerator.

Sources
1. Plomin, R. (1990). *Nature and nurture. Pacific Grove, CA: Brooks/Cole.*
2. Grier, J. W. (1984). *Biology of animal behavior. St. Louis, MO: Times Mirror/Mosby.*

Sexual Wiring of Women's Breasts
Neuroscientists establish breasts as sexual organs

If men have sex on the brain, they are not alone. Recent research found that women's sensory cortex has three distinct areas corresponding to the stimulation of the clitoris, vagina, and cervix[1]. To their surprise, researchers found that self-stimulation of the nipples lights up the same areas. This sheds further light on the sexual importance of breasts.

I have previously addressed some evolutionary reasons for men's fascination with women's breasts and pointed out that stimulation of the breast plays a key role in women's sexual arousal and satisfaction.

The Background
The permanently enlarged human breast is a peculiarity of our species[2]. It may have some signaling value in communicating fertility and plays a role in physical attractiveness.

Breasts are less eroticized in subsistence societies, where women go topless, than in our own where they are exploited in advertising, and in pornography. Even in subsistence societies, breasts are not entirely lacking in sexual significance and are generally stimulated in foreplay according to ethnographic accounts[3].

Moreover, the breasts play a key role in female sexual arousal and we are beginning to understand why in terms of hormones and neuroscience. In their classic report on the female sexual response, Masters and Johnson[4] pointed out that breast volume increases during sexual arousal in addition to changes in the areola and erection of the nipples.

The breast and bonding
The function of the breast in sexual behavior is sometimes attributed to face-to-face copulation that is unusual among mammals. If the breast

is already used for mother-infant bonding, the argument goes, then it is a small step for it to be used in facilitating bonding between lovers. After all, it is in easy reach.

The stimulation of the nipple during breastfeeding increases the amount of the hormone oxytocin that circulates. Oxytocin is often referred to as the "cuddling hormone" because it is released by male and female mammals during close social encounters of various kinds[5].

In addition to its general social effects, whereby a mother feels closeness for the baby she is feeding (and vice versa), there are other more specialized functions of oxytocin. One such function is milk flow - a reflex known as the "milk let-down response" familiar to mothers and dairy farmers alike.

Another is sexual arousal and orgasm. Some women experience intense pleasure, even orgasm, from breastfeeding. This phenomenon was long written off as a mere oddity, but neuroscientists are beginning to understand why it happens.

Sexual wiring of women's brains

The great complexity of the female sexual response may be attributable to the fact that there is not one, but three sensory maps in the parietal cortex that light up in functional MRI images when the genitals are (self) stimulated. One represents the clitoris, another the vagina and the third represents the cervix.

All three of these maps also receive input when the nipple is stimulated. From a functional perspective, this means that the breast doubles as a truly sexual organ. It is not just an exciting visual stimulus for (most) men but also a key source of sexual pleasure for most women. As to the wiring of men's nipples, the jury is out. Some men's nipples are also responsive to sexual stimulation, but the brain response has yet to be mapped.

Sources

1. Komisaruk, B. R., et al. (2011). Women's clitoris, vagina, and cervix mapped on the sensory cortex: fMRI evidence. The Journal of Sexual Medicine, 8, 2822-2830.
2. Barber, N. (1995). The evolutionary psychology of physical attractiveness: Sexual selection and human morphology. Ethology and Sociobiology, 16, 395-424.

Why Liberal Hearts Bleed and Conservatives Don't

Why politicians have trouble listening to each other

Political liberals are "bleeding hearts" because they empathize so strongly with the sufferings of others. As Bill Clinton so succinctly phrased it, "I feel your pain." When Republicans wanted to compete in the empathy department, they had to invent new terminology to identify this strange bird. They called it a "compassionate conservative."

One might ask why conservatives have, or are perceived as having, too little empathy. Why do liberals have too much? A widely credited explanation is in terms of competing world views.

The conservative world view

Conservatives see the world as a challenging place in which there is always someone else who is ready to steal your lunch. Confronted by a potentially hostile environment, the best course is to take precautions to ensure your own well-being and that of your family.

This precautionary stance helps to explain many of the distinctive traits of conservative Americans, as well as right-wing politicians the world over. These traits may reflect a proactive element, such as fighting to be winners in a competitive world, or they can be defensive, involving measures to neutralize specific threats.

The threatening world view illuminates the conservative take on specific political issues in fairly obvious ways.

- Conservatives are pro-gun because they want to be able to defend themselves against criminal threats of any type.

- They are mostly religious because religious rituals foster feelings of safety in a dangerous world such that the most dangerous countries in the world are also the most religious[1].

35

- They tend to be more hostile to immigrants, foreigners, and racial or ethnic minorities and to view them as more of a threat.

- They fear attacks by other nations and therefore support a strong military and a bellicose foreign policy on the theory that a good attack is the best defense.

- Apart from military defense, where government is an asset, conservatives fear government intrusions into their lives and particularly fear having their wealth eroded by taxation.

- They are pro-family because being surrounded by close relatives is the best defense against threats that surround them.

- They oppose welfare for the poor in that it encourages dependence, making the failures of society parasites on the successful, as well as inverting the proper incentive structure.

- They admire wealth because successful people are seen as having worked hard in pursuing a moral obligation to provide for themselves and their families in a difficult and uncertain world.

The liberal world view

The liberal world view is mostly the opposite. Liberals take a more optimistic view of the world regarding it as somewhat more benign. Government is a vehicle through which the citizens of a democracy can solve problems and improve the well-being and happiness of most people.

- Liberals feel that protection of citizens against crime is better left to police and that armed citizens are a threat to those around them.

- They are less religious than conservatives because they perceive the world as less threatening. Moreover, they rely more on science, and education, as a means to solve problems.

- Liberals are more welcoming to immigrants. They are less likely to view foreigners and racial or ethnic minorities as a threat.

- They favor negotiation and consensus-building over warfare in foreign policy and do not believe in excessive military buildups that drain social spending.

- Liberals are happy to pay their taxes if they believe that the money is being used to improve the quality of life of others, whether they are poor or rich. Instead of being a threat, the government reflects the will of the people.

- Liberals are less interested in family ties as a protective bubble.

- They support welfare programs for the poor because these may reduce child poverty, as well as reducing crime and social problems.

- Liberals are suspicious of wealth, feeling that much of it is inherited or obtained through sharp business practices or outright corruption. They also feel that concentrating resources in the hands of the one percent impoverishes everyone else, thereby undermining social trust[1].

Conservatives see the world as a more threatening place because their brains predispose them to being fearful[2]. They are also predisposed by brain biology to hating complexity and compromise. That would help to explain why politics can be so polarized, particularly in a rather conservative era like the present.

It also explains many of the quirky differences between Democrats and Republicans. My favorite is the mind-blowing fact that four times as many Republicans as Democrats have mudrooms in their homes[3]. Got to protect your home from that contaminating mud!

Sources

1. Barber, N. (2012). Why atheism will replace religion: The triumph of earthly pleasures over pie in the sky. E-book, available at:

http://www.amazon.com/Atheism-Will-Replace-Religion-ebook/dp/B00886ZSJ6/

2. Barber, N. (2011, April 19). Conservatives big on fear, brain study finds. Blog post accessed at:

http://www.psychologytoday.com/blog/the-human-beast/201104/conservative...

3. Garchick, L. (2000, November 5). Remodelers split along party lines. San Francisco Chronicle, accessed at:

http://articles.chicagotribune.com/2000-11-05/news/0011050380_1_remodel...

Does Trump Suffer from Narcissistic Personality Disorder?

Narcissism is a personality dimension as well as a clinical diagnosis

Whether narcissism is a real disorder, as opposed to a dimension of personality on which we all vary, is controversial. It may be both. Does Donald Trump conform to the clinical pattern?

Professional psychiatrists and psychotherapists are loath to go on record as saying Trump has a psychiatric disorder on the premise that one cannot give a diagnosis without an office visit. Besides, most narcissists are quite unlikely to recognize that they have a problem and to schedule an appointment. Fortunately, the DSM is written so clearly, and so simply, that the diagnosis is transparent. Here are the symptoms.

Does Trump Have Narcissistic Personality Disorder (NPD)?

According to DSM-5, individuals with NPD have most (at least five) or all of the symptoms listed below (generally without commensurate qualities or accomplishments).

1. Grandiosity with expectations of superior treatment by others.

2. Fixation on fantasies of power, success, intelligence, attractiveness.

3. Self-perception of being unique, superior, and associated with high-status people and institutions.

4. Needing constant admiration from others.

5. Sense of entitlement to special treatment and to obedience from others.

6. Exploitative of others to achieve personal gain.

7. Unwilling to empathize with others' feelings, wishes, or needs.

8. Intensely jealous of others and the belief that others are equally jealous of them.

9. Pompous and arrogant demeanor.

Among other criteria, the symptoms must be severe enough to impair the individual's ability to develop meaningful relationships with others and reduce an individual's ability to function at work. As far as the first of these is concerned, Trump evidently has no close personal friends.

Work function is also an issue. The ghost author of Art of the Deal, Tony Schwartz, found it impossible to interview Trump who quickly became bored. He gleaned most of the necessary information by being a fly on the wall in Trump's office.

Some of the DSM criteria are less relevant to Trump given his birth to money. His life as a plutocrat guarantees contact with high-status persons and being fawned over as a VIP. For those that are clearly relevant, he checks out on all symptoms, it seems. According to DSM criteria, Donald Trump suffers from narcissistic personality disorder.

Can a Narcissist Function as a U.S. President?

It is, perhaps, no surprise that widely held impressions about Trump's narcissism are corroborated by the DSM criteria. The key question to ask is whether, having come so far despite his psychiatric disorder, Trump or any other narcissistic personality can communicate well enough to be an effective leader of the free world. There have been many narcissistic heads of state before but the clearest examples, such as Fidel Castro, Saddam Hussein, and Hugo Chavez, have been dictators.

Narcissists are difficult to deal with, whether as friends or as politicians. They do not feel the need to build consensus, which is why most are screened out by democratic systems of government. We need to be wary of the ones that get through.

Do Lie Detectors Work?

Should you ever take a polygraph?

The term "lie-detector" suggests that a polygraph machine can detect lies. However, this device, which is commonly used in criminal investigations, actually measures nervous excitement. It operates on the premise that if a person is telling the truth, they will remain calm.

Trial by ordeal

The lie detector can be considered a modern variant of the old technique of trial by ordeal. A suspected witch was thrown into a raging river on the premise that if she floated, she was harnessing demonic powers.

Such techniques never had much credibility. A subject who passed the test had a reasonably high probability of drowning that was at least preferable to getting burned at the stake as a witch.

No one dies of a polygraph test, of course, and the results are mostly inadmissible in courts of law. Yet, a person who is otherwise exonerated by the evidence may endure the legal jeopardy of being unfairly incriminated in the eyes of investigators.

Although polygraph machines look scientific and measure responses such as sweating and increased pulse rate with exquisite accuracy, they are crude in their conception. Indeed, they are no more sophisticated than an ancient Arab ordeal for detecting liars.

In the Arab test, a heated knife blade was pressed to the subject's tongue. If he was telling the truth, his tongue would not get burned. The idea is that when people are nervously excited, their mouth goes dry because nervousness suppresses salivation. In principle, the lie detection system involved is exactly the same as for a polygraph test.

Does the polygraph work?

There has been a lot of controversy about whether lie detectors work. Some experts claimed that a high proportion of persons who "failed" the polygraph subsequently confessed to crimes. On the other hand, the test generates a lot of false positives, i.e., people who are telling the truth but whose polygraph test suggests they are lying.

Whereas the American Polygraph Association claims accuracy rates of over 90 percent, leading critics such as David Lykken[1] put the polygraph accuracy rate at around 65 percent. This is only slightly better than the 50 percent correct one would get by flipping a coin.

Interestingly, the polygraph is quite good at identifying liars but does no better than chance at detecting honest people, according to Lykken. In other words, there is a 50:50 chance that a polygraph test will say an honest person is lying (a 50 percent "false positive" rate).

Some researchers have looked for more direct evidence of lying through the analysis of brain scans. Another major weakness is that the test can be faked.

In the normal administration of the test, technicians rely upon responses that they know to be true to provide a baseline against which deceptive answers can be judged as an increase in nervous arousal. One of the most effective means of faking the test is to enhance arousal accompanying honest answers, so that it is difficult to detect increased arousal accompanying lies.

The fact that the test can be fooled in this way also highlights the subjectivity with which examiners judge the results, for there is little standardization of procedures as to how much of a polygraph change indicates lying.

Doubts about polygraph tests grew in the scientific community until the National Research Council – an organization of scientists –

conducted a systematic evaluation and concluded that the test is lacking in scientific validity[2].

In 1998, the U.S. Supreme court acted to restrict their use in legal proceedings. In particular, defense attorneys can no longer use evidence that their client passed a polygraph test as establishing innocence of a crime.

Even as the polygraph test is discredited in legal proceedings, its use also declined in other settings and most employers are legally barred from using it as a technique to recruit honest employees. The government is one exception.

From a scientific perspective, there is no rationale for administering a polygraph test. So, there is certainly no good reason to take one - if you can avoid it. Otherwise, you expose yourself to the nightmare of false self-incrimination.

Sources
1. *Lykken, D. (1984). Polygraphic interrogation. Nature, 307, 681-684.*
2. *Adelson, R. (2004). The polygraph in doubt. APA Monitor, 35, 71.*

The Three Reasons for Polygamy
Something Else We Can Learn From Birds

Both candidates for the presidency in 2008 owed their very existence to polygamy[1]. President Obama's father belonged to the polygamous Luo tribe. Mitt Romney's paternal great grandfathers moved to Mexico to continue the Mormon practice of polygamy then outlawed in the U.S. The time is ripe to ask what advantages polygamy has over monogamy.

Although plural marriage is banned in developed countries, it is surprisingly common and popular elsewhere, with 55 percent of women sharing their husbands in Benin and an average of 16 percent of women doing so in less developed nations[2]. Polygamy may be detested in developed countries, but it is practiced to some degree in most societies studied by anthropologists. What did polygamy do for the Obamas and the Romneys that they could not accomplish with monogamy?

Studies in animal behavior show that polygynous mating systems (i.e., one male mating with several females) have at least three possible advantages.

Polygamy: A bird's eye view
There are three basic reasons for polygyny in birds. First, there may be a scarcity of adult males. Second, some males may have much better genes than others which is particularly important for populations where there is a heavy load of diseases and parasites to which resistance is genetically heritable. Third, females do better by sharing a mate who defends a good territory (with plenty of food and cover) than they would by opting to be the single mate in a bad territory.

So much for birds! Do humans choose polygamy for similar reasons?
My research on 32 countries where polygamy is practiced by at least five percent of married women yielded answers [2]. Polygamy increased

where there was a scarcity of males in the population (first reason for birds).

Countries having a heavy infectious disease load had many more polygamous marriages (second reason for birds.) Women in disease-prone countries may prefer highly disease-resistant (i.e., physically attractive) men to father their offspring leaving less desirable men without mates. There is independent evidence that women care more about physical attractiveness in these countries and have a higher sex drive[3].

Having economic resources facilitates polygamy for humans consistent with resource-defense polygyny in birds (reason number 3). Thus, there were more polygamous wives in countries where men could monopolize wealth whether in terms of earned income or farmland (analogous to animal territories.) My findings were not new: they corroborated earlier research but used better data.

So, humans turn to multiple marriage for the same three basic reasons that birds do (scarcity of males, selection for disease-resistant genes, and defense of breeding territory and its economic equivalents.)

In my study, I also evaluated a number of "explanations" for polygamy that are routinely trotted out by social scientists and other observers in developed countries who find polygamy repulsive.

Contrary to popular assumptions, multiple marriage had nothing to do with poverty, backwardness, or oppression of women (e.g., acceptance of wife-beating) in my study. Of course, that begs the question as to why polygamy survives mostly in underdeveloped countries close to the equator and why it is so unpopular in developed countries

Why the developed world hates polygamy
At least three factors are critical. First, instead of a scarcity of males, developed countries have an excess, thanks to better public health that saves more males than females. Second, colder winters made it impossible

historically for mothers to raise children without substantial help from their husbands.

The most important reason that polygamy is out of place in the modern world is that it works best in agricultural societies where children contribute to farm labor and care of livestock[4].

Developed countries are highly urbanized and it is very difficult to raise large families in cities because children are a huge drain on finances that lasts for two decades thanks to the extent of modern education. In agricultural societies, by contrast, children defray the expense of raising them by contributing productive labor to the household economy.

Neither Romney nor Obama had any desire to discuss their polygamous background. Both are religious but they conveniently forgot the beloved patriarchs of the Old Testament like David, Solomon, and Abraham and their many wives (which neighbors were told not to covet).

Romney's Mormon ancestors practiced polygamy but it was mainly confined to members of the church hierarchy who were wealthier than others in terms of land holdings and could maintain multiple households (bird reason number three).

Obama's Luo ancestors likely practiced polygamy for all three bird reasons. There was a scarcity of males, local diseases were a major issue and powerful men could monopolize wealth.

The fact that both candidates were descended from recent polygamous ancestors[1] when no other presidential candidate ever was is a remarkable instance of American diversity. It should be cherished rather than otherwise.

Sources

1. Maraniss, D. (2012, April 12). *The polygamists in Obama and Romney's family trees. The Washington Post.* http://www.washingtonpost.com/opinions/obama-and-romney-both-come-from-...

2. Barber, N. (2008). *Explaining cross-national differences in polygyny intensity. Cross- Cultural Research, 42, 103-117.*

3. Barber, N. (2008). *Cross-cultural variation in the motivation for uncommitted sex: The role of disease and social risks. Evolutionary Psychology, 6(2): 217-228.*

4. Barber, N. (2009). *The wide world of polygamy: We hate it, others love it. Blog post. Psychology Today.* http://www.psychologytoday.com/blog/the-human-beast/200902/the-wide-world-web.

Why People Cheat
Sleeping around is more of a mystery than ever

Cheating is a fact of life. Ostensibly monogamous relationships in the animal world are not really monogamous. About a quarter of the offspring raised in nests of socially monogamous birds, such as barn swallows, are sired by males other than the mate.

I have previously argued that infidelity is a part of the human heritage also. But why? If a man, or woman, is unhappy with some aspect of their live-in lover, or spouse, why did they agree to shack up with them in the first place?

Not long ago, this would have seemed an easy problem to researchers. For women, the main benefit of infidelity was better genes for her offspring with the ancillary benefit of gifts provided by lovers. For a man, the advantage was simply spreading his genes around or siring more children.

Goodbye wild oats theory

In recent years, both these theories have been dashed on the rocks of skepticism and contradictory evidence. As to the male explanation, this works only if men differ widely from each other in the number of children they sire. Yet, recent data on subsistence human societies contradict this view: men's reproductive success is no more variable than that of women[1]. The bottom line is that seeking to maximize the number of children sired was not a viable (unconscious) reproductive strategy for men. One plausible reason is that fathers were much more important in providing nutrition for young children than anthropologists used to think.

The female rationale for infidelity was simpler. A woman who threw in her lot with the best man available had second thoughts when she beheld the sculpted figure of a natural athlete. She yielded to temptation and her progeny benefited from having better genes.

Goodbye good genes theory

The key problem with the good genes theory is that no one has ever been able to nail down exactly what makes genes good. If there is plenty of food, it is fine to look like Mr. Universe. Most local ecologies, however, would be far too poor to support that sort of physique and a much slighter build is found for men in most subsistence societies.

Any gene that is good for one environment may prove a liability if the environment changes. Perhaps that is why the evidence relevant to good genes is so bafflingly inconclusive as I pointed out in my post on the peacock's tale.

Even if there really were good genes, and if females preferred to mate with males having these good genes, then the bad alternatives would get bred out and all males would have to have good variants.

Parasite resistance

One aspect of the good genes theory that seems to hold up in animal research is the view that females select males who are resistant to local diseases and parasites.

This theory seems to hold up for women also. In societies where parasites are prevalent, women are more interested in physical attractiveness of a man. They are also more interested in casual sex, suggesting that one motive for infidelity is to acquire better immunity for their children.

Another fairly obvious benefit from infidelity for females is to look for a new permanent partner. Women generally engage in extramarital relationships when they are dissatisfied with the emotional quality of their marriage.

Sauce for the gander

So much for women! Why do modern men engage in extramarital sex if this was not particularly likely to increase the number of offspring raised to maturity by their ancestors? There is little evidence that dissatisfaction with their spouse motivates men to cheat as a way of initiating a new marriage.

Men may be pursued by women intending to supplant their spouse, of course. Otherwise, there are two plausible masculine motives for cheating - the desire for sexual variety, and attraction to younger, more fertile, women. The first seems like reproductive opportunism but the second might reflect the psychology behind selection of a long-term mate.

Men are interested in sex without strings attached but cheating alone is not a viable reproductive strategy. The reason is that the vast majority of women resist it.

Source
1. Brown, J., Laland, K. N., & Borgerhoff Mulder, M. (2009). Bateman's principles and human sex roles. Trends in Ecology and Evolution, 24, 297-304.

Dead at 27: Why Highly Creative People Die Young
Young falling stars evoke powerful emotions

When poets and rock stars die young, we may attribute their tragic early loss to a pattern, such as the seemingly cursed age of 27 years. This was the lifespan of Jim Morrison, Janis Joplin, Brian Jones, Jimi Hendrix, Kurt Cobain, and Amy Winehouse.

This age may be a real limit, or it may be the result of a confirmation bias, where confirming cases are collected, as in the documentary on short-lived prominent musicians (27: Gone Too Soon).

Possible factors include suicidal behavior, drug abuse, sex hormones, and the sense of career failure.

Suicide

Many premature deaths are linked to suicidal behavior. The issue is not that the individual necessarily intends to kill themselves. Yet, they take great risks unnecessarily, whether it is driving while intoxicated, taking out a boat at night in rough seas, or consuming many different recreational drugs in high doses.

While many creative people are high in risk-taking, suicide statistics shed doubt on this explanation. The median age of suicide is not 27 but 39. Young people have generally low suicide rates in comparison to older cohorts with rates being highest for those over 60.

Drug Abuse

If suicide is not a plausible explanation for tragic deaths among young creative people, it is possible that persistent drug abuse has the effect of prematurely aging their internal organs, making longevity impossible.

Being a prominent figure in popular music is not an easy career path. To begin with, many experience the pressures of failure and use recreational drugs as an escape mechanism. Then, there are even the greater pressures of success and fame.

Habitual use of recreational drugs exacts a toll on the internal organs. If this pattern begins in the teenage years, it has the potential to bring on vital organ failure, and premature death within a decade.

For this reason, alcoholics who begin heavy drinking in their teens fail to reach their 30s. Many entertainers use alcohol, in addition to other drugs, to get through their day. Organ toxicity is one plausible reason for the pattern of death by 27 years.

Yet, there is variability in individual physiological responses to drugs of abuse. While deaths might well cluster at age 27, there is no obvious reason that it would not be 25 or 29. The mid-20s is a time of physiological changes for people who are not drug addicts. For men, it is the peak time of testosterone production. This is linked to risk-taking and accidental death[1].

Sex Hormones and Creativity

Most of the extremely early deaths of entertainers are in men. Young men are particularly vulnerable to accidental death and this implicates sex hormones[1]. Testosterone peaks in the mid-20s and testosterone is predictive of risk-taking, accidents, and violent death.

This age is also a peak of male artistic creativity and individual accomplishment in various fields from sports to mathematics[2]. The same is true of entertainment professionals.

The sexual exploits of male rock stars are legendary and their sexual magnetism draws an endless supply of enthusiastic female partners. Such excesses go along with drug abuse, irrational risk-taking, and high levels of accidental death, whether from risky driving, accidental drowning, or accidents. Such behavior carries a price and is associated with very early deaths.

Female entertainers have no particular physiological vulnerability in the mid-20s. Yet, they are encountering the peak of physical attractiveness that challenges entertainers.

Feelings of Failure

The pressures of career success strike early because most creatives make their mark before the age of 30. This general principle is as true of scientists and mathematicians as it is of poets and musicians.

There are some exceptions, of course. Novelists may not hit their stride until their 40s and some painters and poets produce their most influential work in old age.

Even so, the three-decade cutoff looms. Those who have not achieved their ambitions by the age of 27 may feel that time is running out. Entertainers feel more intense pressure to succeed than people in more humdrum occupations thanks to the cut-throat nature of their business. Stress may reduce life expectancy.

Stress and Poor Health Behavior

Despite the fraction of entertainers who flame out tragically early, most entertainers experience old age.

On average, entertainers survive for just a few years shorter than others, according to a study of New York Times obituaries. This deficit has two likely causes. The first is a greater incidence of smoking—based on the number of deaths to lung cancer. The other is psychological stress.

Most professional entertainers experience financial privation with spells of abject failure. Even the successful ones worry about getting pushed aside by new talent.

What It All Means

The phenomenon of dying at 27 could represent the physiological limits of a stressful life compounded by heavy drug abuse. Yet, we cannot be confident that more entertainers die at 27 than at 28, or 26. Most entertainers live far longer.

The age of 27 years may be selected because it is maximally tragic. After all, the individual has reached the acme of their talent and already climbed a mountain of success. It is a great height to fall from.

Sources

1. Courtenay, W. H. (2000). *Behavioral factors associated with disease, injury, and death, among men: Evidence and implications for prevention. Journal of Men's Studies, 9, 81-142.*
2. Miller, G. F. (1999). *Sexual selection for cultural display. In R. Dunbar, C. Knight, & C. Power (Eds.), The evolution of culture: An interdisciplinary view (pp.71-91). New Brunswick: Rutgers University Press.*

Why Do Some Poor People Vote Against Their Interests?

If head and heart are at odds, emotion often wins out

W hen poor people put their faith in conservative leadership, one might ask why they trust representatives of a wealthy owner class that will likely never include them. Yet there is a rationale based on the emotion of fear.

The Psychology of Living in a Dangerous World

When young mammals mature in a place with high exposure to predators, they adjust their behavior. They spend more time in safe places, such as burrows, and less time wandering about above ground. With less time exploring their environment they have fewer opportunities for brain-enriching experiences. As a result, they are less good at adapting to change[1].

Lab researchers found that these effects were mediated by stress hormones associated with frightening experiences such as the approach of a predator.

Most children do not have to worry about predators, but they vary greatly both in the objective realities of their lives, such as the use of corporal punishment by parents, and their subjective responses to frightening experiences. They are also influenced by adult perceptions of danger in their local community.

Conservatism and Fearfulness

Political conservatives (defined as high scorers on a right-wing authoritarianism scale) experience fear more intensely. This propensity is related to brain anatomy and physiology. Sensitivity to fear probably reflects a combination of influences from genetics and childhood experiences.

Whatever the causes, signs of conservative leanings are present early in childhood before children are engaged in political issues. Children who are sticklers for the rules in games with other children likely go on to vote for conservative leaders[2,3]. In other respects, they tend to be rather rigid in their behavior and find it difficult to make new friends.

Such fear of the unpredictable reflects a sensitivity to danger mediated by limbic-system activation. This profile probably reflects mammalian adaptations to actual risks in the environment[4].

Children growing up in extreme poverty, or children exposed to abusive parents, also grow up believing, for very good reason, that their lives are risky and that caution is warranted.

Poverty and Insecurity

If conservatives generally believe that the world is a dangerous place regardless of their individual experiences, those raised in poverty have a very good reason for the same belief rooted in their own lives.

Poverty often implies greater health problems, violence, the early death of a close relative, high crime risk, food insecurity, drug addiction, or lack of adequate health care.

Belief in a dangerous world is connected to various conservative policies. A strong military is supported to counter international threats. Because there are very bad criminals out there, there is a need for severe penalties, up to a death sentence, to keep them out of circulation.

Similarly, corporal punishment is needed to socialize children in obedience to authority. Just as other nations harbor a great deal of ill will, immigrants must be treated with suspicion and held at arm's length as potential sources of crime and disease. It is important to amass as much wealth as possible because the future is uncertain and you cannot rely on the government to solve your economic problems in a dog-eat-dog world.

To the extent that the fear-based sensibility of poor people overlaps with that of conservatives, we can expect their political views to coincide also. This means that playing on popular fears and ethnic tensions is good for conservatives in elections.

According to a conservative sensibility, the only source of reliable aid and support in difficult times is our own family. So, we must respect our elders and do everything we can to honor them and preserve their traditions, including their religious beliefs.

The Religious Nexus

Just as there is a marked intersection between the emotive aspects of conservatism and those of being raised under stressful conditions, there is also an overlap between both and religion. One way of describing this connection is to think about religion as a mechanism for coping with fear and uncertainty about what the future holds as developed elsewhere in this book under "Why Atheism Will Replace Religion."

The central idea is that as countries develop, people enjoy a better quality of life with improved health and life expectancy. They are also more secure about what the future holds for them (i.e., they have existential security). In societies like our own, that are bedeviled by sharp income inequality, there is less existential security and religion has a stronger hold.

Along with love of family and tradition, and relative lack of openness to new people and ideas, conservatives focus on religion as a means of preserving their way of life and resisting change. It is also a way of drawing in poor people and thereby inducing them to vote for policies that go against their economic interests or otherwise reduce their quality of life. For example, many poorer Americans voted for a party that promised to take away their health care.

How does one get people to vote against themselves? The answer is mostly by an appeal to various kinds of fear, including the fear of God.

Economic Self-Interest

Conservative leaders must convince followers of two things. First, the world we live in is full of threats. Second, supporting that leader is the only way of protecting themselves from the threats.

If the first goal is achieved, the second is relatively simple. After all, if some politician is the only one who emphasized a specific threat, then it makes sense that they would be the only one with an answer. The list of potential threats is long, ranging from foreign military threats to domestic terrorism, exaggerated fears of minorities and immigrants, pluralism, disease, violent crime, or government overreach.

Conservatives also play on fears that their religion is under threat. This ploy succeeded in cases as different as the US South and Putin's Russia. If religion is a bastion against many dangers, then anything that weakens it is threatening to the poor.

Conservative politicians who are otherwise highly secular in their behavior and sensibility fake piety in order to get elected and promote the causes of right-wing religious extremists to maintain support. Such tactics are highly effective and may induce poor people to vote against their economic self-interest and in favor of a wealthy elite that becomes wealthier by exploiting them.

Sources

1. Rosenzweig, M. R. (1996). *Aspects of the search for neural mechanisms of memory. Annual Review of Psychology*, 47, 1-33.
2. Tuschman, A. (2013). *Our political nature: The evolutionary origins of what divides us*. Amherst, NY: Prometheus.
3. Garcia, H. A. (2019). *Sex, power, and partisanship: How evolutionary science makes sense of our political divide*. Amherst, NY: Prometheus Books.
4. Kalinichev, M., Easterling, K. W., Plotsky, P. M., and Holtzman, S. G. (2002). *Long-lasting changes in stress-induced corticosterone response and anxiety-like behaviors as a consequence of neonatal maternal separation in Long-Evans rats. Pharmacology, Biochemistry, and Behavior*, 73, 131-140.

Gender Differences of Animals
Sex differences are real, biologically based, and here to stay

In an era of increasing gender fluidity, it is worth asking whether gender differences in psychology have been overstated in the past. One way of assessing this is in terms of gender differences of other species that lack our political agendas.

Size, Strength, and Morphology

Among mammals, males are generally larger, and more aggressive than females. This pattern is related to greater production of testosterone that is also associated with increased aggression.

There are intriguing exceptions including hyenas, and chinchillas, where the females are larger and more socially dominant.

Such exceptions are sometimes related to increased testosterone production by females. This involves activity in the adrenal gland that is the primary source of female testosterone.

Hyenas are particularly interesting because, in addition to being larger than males, females have a pseudo penis superficially indistinguishable from that of males that plays a role in social greeting and dominance relationships.

Size affects behavior, giving larger individuals an advantage in establishing social dominance. This means that females of most species defer to males.

In addition to being larger, males generally are the ones who develop bodily signals designed to facilitate reproductive success by attracting females and intimidating rivals. Examples include the antlers of deer and the beards of humans.

Human females are unusual in carrying most of the sexually attractive signaling from youthful facial features to permanently enlarged breasts. This may be because men invest far more in offspring than most other male mammals do so that they are in greater demand as mates. Females thus bear the burden of advertisement.

If males are generally larger, there are other behavioral traits that generally distinguish the sexes.

Aggression and Risk-Taking

Among humans, men are responsible for about nine-tenths of serious crimes of violence, although women are more verbally aggressive and more willing to pick a fight.

In addition to being less involved in dangerous physical aggression, women are generally more risk averse. They avoid situations that are threatening to life and limb. For example, there are few societies where women participate in warfare.

If aggression can be considered a masculine adaptation for mating competition, risk aversion of females makes sense given their greater investment in children. They may be particularly risk-averse if they are mothers of young children given that their injury, or possible death, would have adverse consequences for their children in terms of survival and social success. These patterns likely reflect the psychological differences of other mammals.

Parental behavior itself is another conspicuous sex difference among mammals although different species vary greatly in the extent of paternal investment in offspring.

Parental Behavior

Sex differences in parental behavior are generally broad-ranging, early-developing, and consequential. As the gender with the largest investment in offspring, females are more strongly disposed to make that investment. Beginning with a larger gamete, the egg, females bear the huge additional costs of gestation and lactation.

Among mammals, as a group, females are more strongly inclined to care for young than males. Indeed, males of many species are hostile to young and may kill them on sight as a way of bringing the mother into reproductive condition earlier. This occurs among several primates, and even in human indigenous societies such as the Ache of Paraguay[1].

There are some intriguing exceptions, such as lion tamarins where males do most of the care of young.

Of course, many men are now the primary caregivers for children. Interestingly, new fathers experience a decline in testosterone levels as they adjust to a nurturing way of life.

Another arena where the reality of gender differences in biology strikes home is in health.

Health

Women and men are each more vulnerable to specific diseases than the other. For example, women are more vulnerable to rheumatoid arthritis whereas men have higher rates of heart disease. To the extent that these differences are a feature of biological differences, they are presumably present in other species.

Such differences are sometimes explainable in terms of varied hormone profiles. For example, testosterone suppresses immune function, which could explain why women are more vulnerable to autoimmune disorders like rheumatoid arthritis.

Women enjoy a significant health advantage over men in developed countries and enjoy life expectancies that are about four to five years longer.

This health advantage is complex and involves more than hormones. Genes are expressed differently in males and females, which can have health consequences. More important is the fact that women have better health behavior. They tend to take fewer health risks and take better care of their health. They are also better at avoiding violence and are less likely to die through accidents.

Another intriguing mechanism is the fact that caregivers generally live longer with primate males living as long as females if they care for offspring. When men take the primary role in caring for children, their testosterone level falls, suggesting one mechanism through which the influence of caring on health may be expressed.

Males are generally socially dominant over female mammals but there are exceptions.

Status and Leadership

In general, if females are larger than males, they tend to be socially dominant, as is true of chinchillas and hyenas, for example. This phenomenon is related to social leadership among primates where a single dominant male is a common pattern.

Some mammalian societies are matriarchies, however. This is true of elephants where senior females lead groups across long migrations to food and water. In this case, acquired knowledge is more important than size, strength, or aggression.

This is clearly true of modern societies where women are slowly acquiring an equal position in every area of leadership.

Source

1. Hill, K., and Hurtado, M. (1996). *Ache life history*. New York: Aldine de Gruyter.

The Resurgence of Nationalism
Why is ethnic nationalism rising now?

As a rule, when countries become more affluent, their residents get more liberal, more open to other groups. In recent history, wealth increased but politics became increasingly conservative and xenophobic. Why did the rule fail?

The Rule

As countries become more developed, there are measurable changes in social attitudes that move away from ethnocentrism in the direction of social tolerance and inclusiveness across a range of topics, from gay marriage and women's rights to humane treatment of prisoners[1].

Painted in broad strokes, wealthier countries are more liberal, whereas poorer ones are more restrictive and less tolerant of diversity. The driver here probably is economics. We can draw this inference because when economic contractions occur, attitudes revert to being less liberal[1].

Why do such shifts occur? Social scientists have not offered any good causal explanation, but a Darwinian approach to animal behavior offers useful perspective.

Darwinian Competition?

Mammals growing up in a stressful environment differ from those inhabiting a more secure habitat where food is plentiful and predators are scarce.

One consequence of more competitive environments is that mammals are more fearful and hostile, less trusting of others. Another is that they are less open to taking the risk of exploring their environment.

The implications for humans are quite profound, ranging from academic under- performance and academic failure, to troubled personal relationships and increased risk of crime[2].

There are possible political ramifications given that political conservatives are more strongly motivated by fear, including fear of others.

Interesting as such developmental phenomena are, they cannot account for the rapid shifts in political attitudes due to recessions that typically recur on a time scale of less than a decade. It is nevertheless reasonable to assume that human behavior, and that of other species, responds adaptively to changing social conditions. When times get tougher, people get tougher too.

Can this reasoning account for the resurgence of nationalism and extreme right-wing opinions? There are several reasons that the lives of ordinary people have become more stressful, accentuating their shift to the right even in a period of unprecedented affluence.

Income Distribution

One obvious problem in the US and some other affluent countries is that, in a period of increasing financial wealth, the share of income going to the lower half of the distribution is not rising. This generates unhappiness and disillusionment. We must not forget that the extreme right-wing movements of Europe in the 1930's were bred of the Great Depression and the crushed dreams of working people, who were offered ethnic minorities as scapegoats for their problems.

Financial despair gets expressed in drug addictions. One thinks of soaring alcoholism rates in Russia following the collapse of the Soviet Union or the epidemic of addiction to opiates, crack cocaine, and amphetamines today, which are worst in economically depressed districts.

Perhaps the biggest source of a shift to the right is the sense that our society is crumbling and needs to be restored—that our lives are haunted by insecurity and crime.

Low Social Cohesion

Trust in other people is low in the US compared to other developed countries[3]. This phenomenon is associated with higher crime rates and ethnic hostilities.

Social malaise rises from failures of government, signified by collapsing infrastructure, declining school performance, soaring incarceration rates, stubbornly high poverty levels, and so on. Everyone feels vulnerable in such a climate, including political leaders and financial elites.

Many of these problems surfaced in the Great Depression immediately before a rise of nationalist extremism. In our own time, economic and political realities are very different. Also different is the level of our exposure to disturbing news stories, sensational videos, and other paranoia-inducing aspects of modern media.

Mass Media and The Internet

Since the early days of television, psychologists fretted over the possibility that exposure to violent images in entertainment and news would increase violent behavior in the society as a whole. While such fears were exaggerated, it is becoming clear that a diet of sensationalist gore has a rather different outcome.

People are convinced that we live in a very dangerous world, despite the fact that violent crime is lower than it has ever been in history[4]. News media are partly to blame for the distorted view of reality that is propagated with their "if it bleeds, it leads" priorities.

If legitimate journalism may have that effect, people who are exposed to Internet news feeds that filter out balanced reporting in favor of a one-sided political perspective are more vulnerable to paranoid representations of other opinions and ethnicities. That is true even if they do not fall for the fake news that has recently invaded trusted social media platforms.

It does not help that mainstream political leaders are aligning themselves with white supremacist groups and neo-Nazis, contrary to a six-decade tradition of eliminating hate speech from politics.

International Politics

Similar trends are playing out around the globe with the xenophobic British trying to wall off their island from refugees. Then we have the insanity of impending protectionist tariff wars that end badly for everyone. Meanwhile, we have China-first maritime imperialism, Russia-first territorial expansion, and even Canada-first rhetoric.

We have seen this movie before. The previous two versions ended badly. No one should want to see it again.

Sources

1. Inglehart, R., and Welzel, C. (2005). *Modernization, cultural change, and democracy.* Cambridge, UK: Cambridge University Press.
2. Delaney-Black, V., Covington, C., Ondersma, S. J., Nordstrom-Klee, B., Templin, T., Ager, L., et al. (2002). *Violence exposure, trauma, and IQ and/or reading deficits among urban children. Archives of Pediatric and Adolescent Medicine, 156, 280-285.*
3. Zuckerman, P. (2008). *Society without God: What the least religious nations can tell us about contentment.* New York: New York University Press.
4. Pinker, S. (2011). *The better angels of our nature: Why violence has declined.* New York: Viking Penguin.

The Psychology of Insults

The desire to put others down may be as old as...chickens

N ow that we have lived through an election largely fought—and won—on the basis of insults, it is time for an epidemiology of the put-down. What is the underlying psychology of insults - and why do they suddenly seem to be everywhere?

Motivated by Anger?

Chickens are famous for having a pecking order, in which the bottom chicken in the hierarchy is pecked by everyone else and the top chicken is not picked on by anyone. The chicken hierarchy is settled by physical aggression.

In a verbal society, such as the human one, physical aggression is less often used to settle issues of status: these are mostly deferred to verbal interactions. An insult can thus be interpreted as an attempt to reduce the social status of the recipient and raise the relative status of the insulter.

If that logic is correct, we can assume that insults are often motivated by anger surrounding issues of status insecurity. Many insults are reactive: they are responses to real or imagined slights from others, such as a person accidentally cutting in front of someone else in a line.

We live in a period of extreme concern about how we are perceived by others - social psychologists are charting a steady increase in narcissism among college students[1]. There is little consensus about why this is happening, but some scholars believe that the more children are measured on evaluative scales—aptitude tests, IQ scores, and GPA—the more sensitive they are to threats to their social rank.

This narcissism trend is heavily accentuated by social media with participants being subject to unrelenting evaluation by other network members which encourages participants to inflate their egos, often at the expense of others[1]. Concern with how one is perceived creates social insecurity that may be relieved by lashing out at other chickens (or people). Social networks are replete with individuals who deliver stinging rebukes because they enjoy doing so and because they are mostly exempt from the reprisals that one might expect for face-to-face put-downs.

Content: Status, Competence, Sex, and Hygiene

The purpose of a put-down is to reduce someone else in the imagined status hierarchy. It is hardly surprising then, that insults will often refer to a person's social status in terms of ancestry, lack of prestige, or membership in a despised out-group such as Nazis or vagrants, for example.

The content of insults across the ages is monotonously predictable. Many insults feature a sexual component, refer to sexual organs, or bring up shameful or ineffectual sexual behavior. In addition to status and sexuality, insults inflict shame by mentioning unappealing traits—fatness, shortness, baldness, spottiness, and contagious diseases.

Another way of taking a person down is by questioning their intelligence or general mental competence. For insult purposes, recipients are invariably "stupid" or "crazy."

The pecking-order logic of insults means that if the recipient is shamed, then the insulter rises in status relative to the victim. The insulter is the one doing the pecking rather than getting pecked. Not all insults are equal, of course. Some pecks miss their mark and have no impact upon relative status.

Aim: An Arrow Shot Over the House That Hits No One

We live at a time when insults are dispensed so freely that they threaten the financial viability of social media platforms like Twitter. Indeed, Twitter recently issued a code of conduct for users designed to exclude the worst offenders; other sites like Facebook have quickly followed suit.

For people who enjoy dispensing insults, the internet is a perfect environment in that it can offer a shield of anonymity and an absence of consequences. The larger question is whether the frequency of ill-motivated personal attacks will ride the wave of increasing narcissism or succumb to social controls?

Is There a Future in Insults?

Effective communities maintain solidarity by keeping direct insults to a minimum. The elaborate traditions of politeness and respect found in real-world communities of the past meant people behaved in a certain way to avoid unnecessary anger, disputes, and violence.

Online communities are increasingly concerned about the destructive consequences of tolerating the flamers, trolls, and vandals in their midst and are instituting mechanisms for group punishment where those who violate codes of decency are identified and excluded.

Such mechanisms are already ingrained in applications such as Uber and Airbnb. Soon social media will also be regulated - and it is about time. The only problem is that users may ultimately receive an online civility score that will boost collective narcissism - and make us want to peck our neighbors even more.

The chickens are restless!

Source

1. Twenge, J. M., et al. (2009). *Egos inflating over time: A cross-temporal meta-analysis of the narcissistic personality inventory. Journal of Personality, 76, 875-902.*

Are Women More Sexually Faithful Than Men?
Surveys saying women are more faithful defy common sense

We are accustomed to hearing that men are eagerly trying to propagate their genes into future generations via casual sex. If that is true, then one might expect that husbands would be much more likely to have affairs than wives. Are women truly more faithful than men?

Men's adaptations for polygyny

Masculine sexual anatomy and physiology suggest that natural selection designed men for mating with multiple women. The fact that men are taller and heavier than women is indicative of mild polygyny. The idea is that body size is an advantage in fighting over sexual access to females.

Moderately large human testicles also produce a larger volume of semen than is required in a strictly monogamous species, such as gibbons[1]. Men also produce a larger ejaculate upon reunion with their mates following a separation[2]. This suggests sperm competition, or a race by sperms from different men to fertilize an egg.

Men's penchant for casual sex is a commonplace of evolutionary psychology and it creates the market for pornography and prostitution, industries that have few female customers. In addition to these psychological propensities for sex with different women, men report many partners when questioned in surveys.

Women's adaptations for polyandry

If men are adapted for sperm competition, this implies that women must have sometimes had sex with different men during the five-day window that sperm remains viable in the female reproductive tract.

More than any other female mammal, women are equipped to experience sexual pleasure[3]. While the whole-body changes that take place during sexual interactions might strengthen a monogamous bond, they could also motivate women to seek out different sex partners.

Women are said to have continuous sexual receptivity. Unlike other mammals that restrict mating to a few days of the year when they are most likely to conceive, women can have sex at any time[4]. Interestingly, they are more attracted to he-man physiques during the most fertile days of the cycle. This implies that their sexuality is on a dual track. When most fertile, they seek men whose appearance indicates good genes. The rest of the time, they are less selective, and scholars speculate that sexuality is used to extract resources from men.

In theory, women might have affairs to obtain better genes for their offspring. Yet, this happens rarely, and husbands have a high confidence of paternity in most societies[3]. The problem is that infidelity is extremely risky because it can destroy marriages, to the detriment of children, and expose women.

Wives may well be interested in sexual pleasure for its own sake, but casual sex is more costly for them in terms of unwanted pregnancy, reputational damage, and fear of sexual assault. Their opportunities may also be restricted by spending more time caring for children and also by the fact that many women avoid sex while menstruating. Statistically, this means that wives would have fewer affairs than their husbands.

Just the facts

So, husbands evidently have both a higher motivation and more opportunities for casual sex than wives do. Consistent with these notions, women characteristically report about half the lifetime sex partners reported by men. Researchers generally do not believe the difference, as it takes two to tango[5].

If married men are interested in casual sex for its own sake, most do not act upon such impulses, possibly because of the threat to their marriage. Yet, they are twice as likely to cheat as their wives are if survey responses are to be trusted[6].

Whereas men who cheat are motivated primarily by sexual pleasure, women are most likely to have affairs if they are dissatisfied with their marriages[3]. Infidelity is therefore a way of moving on from a bad marriage to one that is more satisfying.

How Differences in Sexual Desire Affect a Marriage
As we see, men may indeed be more likely to cheat because they are motivated by sexual pleasure. This sometimes makes them appear shallow and irresponsible. On the other hand, by being more in control of their sexual impulses and using an affair to transition out of their marriage, women can seem manipulative. No one is at their best while cheating.

Sources
1. Birkhead, T. (2000). *Promiscuity: An evolutionary history of sperm competition.* Cambridge, MA: Harvard University Press.
2. Baker, R. S., and Bellis, M. A. (1995). *Human sperm competition.* London: Chapman & Hall.
3. Barber, N. (2002). *The science of romance.* Buffalo, NY: Prometheus.
4. Thornhill, R., & Gangestad, S. W. (2010). *The evolutionary biology of human female sexuality.* New York: Oxford University Press.
5. Einon, D. (1994). Are men more promiscuous than women? *Ethology and Sociobiology, 15,* 131-143.
6. Wiederman, M. W. (1997). Extramarital sex: Prevalence and correlates in a national survey. *Journal of Sex Research, 34,* 167-174.

Do Humans Need Meat?

Environmentalists encourage us to cut down. Is that a good idea for our health?

Environmentalists encourage us to cut down on meat consumption in favor of vegetable foods that are less damaging to the environment. Given that our ancestors likely had plenty of meat in their diet, is going meatless a good idea?

The History of Meat-Eating

Our chimpanzee-like ancestors were mostly vegetarian, judging from the diet of modern chimpanzees that subsist mainly on fruit, leaves, and nuts, with a rare morsel of hunted meat. After leaving the forests in favor of open grasslands, hominids likely increased the proportion of meat in their diet - they would have encountered large herds of game animals.

Initially, meat was consumed raw. About 200,000 years ago, the first hearths appeared, and there is genetic evidence that the human brain began to burn a great deal of energy[1]. Cooking partially breaks down food, making it easier to digest. Thanks to the culinary arts, the human gut had less work to do and became much smaller than the digestive system of a herbivorous ape.

At this point, it seems that our ancestors were partly specialized as meat-eaters, although they likely continued to eat a wide range of vegetable foods.

With increased energy use in the brain, we suddenly became a lot smarter. Key evidence for this is that our ancestors refined their toolkit into the efficient technology for killing at a distance that drove many large prey species into extinction around the globe (an event known as the Pleistocene overkill). Everywhere humans migrated, the extinction of many large prey animals soon followed.

Assuming that humans were responsible, our forebears must have eaten a great deal of meat. Ultimately, they may have depleted prey animals so much that they were forced into agriculture to avoid starvation[2].

Even today, meat occupies a special place in the diet, being a preferred food in many societies and therefore taking pride of place at celebrations from the Thanksgiving turkey in this country to the pig feasts that the Enga of New Guinea hosted before making war on their enemies[2].

Meat Hunger and Nutritional Deficiency

We can assume that meat was an important component of the diet right up to the Agricultural Revolution when people began to rely heavily on a small number of cereal crops, such as wheat and rice.

The immediate consequence of this dietary shift was a decline in health and life expectancy. Early agriculturalists were shorter in stature and had lower life expectancy compared with their forager ancestors[2]. It seems likely that their health problems were caused more by a decline in nutritional variety than by the loss of meat per se.

There is an ongoing controversy about the adequacy of vegetarian diets. Although vegans—who avoid meat, eggs, and fish—are at risk of nutritional deficiency problems, most experts agree that a wise choice of foods can ameliorate the problems. So, lack of calcium can be addressed by eating collard greens, or tofu, for instance. A scarcity of vitamin B12 can cause anemia and nerve damage but is easily addressed by taking supplements.

In general, modern-day vegetarians are as healthy as their meat-eating counterparts and actually have lower rates of heart disease.

Meat as an Addiction?

Despite limited evidence of the nutritional necessity for meat, people behave very much as though it were a vital component of the diet. A recent book[3] argues that humans are obsessed with meat, noting that in many languages a distinction is made between hunger in general and deprivation of meat.

People who have plenty of vegetables experience "meat hunger." For that reason, African forest peoples who live largely by hunting have trouble accepting a diet dominated by grains and vegetables[4].

People are hooked on meat due to its taste properties that combine umami (a delicious taste also associated with tomatoes), saltiness, and the distinctive taste of seared fats.

Meat hunger is doubtless controlled by the sensory pleasures of eating animal foods. Why are people so obsessed with meat if vegetable foods provide equivalent nutrients? One long-standing theory, developed by anthropologist Marvin Harris, is that people living in a protein-poor environment value meat highly because it is the quickest way for them to secure a balanced diet. Hence the phenomenon of indigenous people, who are well-fed on foods such as bananas, experiencing a powerful sense of meat deprivation.

Instead of hunting large game, they could theoretically look for alternative protein sources, such as nuts, legumes, or mushrooms. The problem is that such foods are characteristically in short supply for much of the year so that hunted food may be a quick fix for deficiencies of protein and other vital nutrients.

Of course, a solution that worked for our remote ancestors may be out of place in a world where the planet is so crowded that meat production is a strain on global resources. In the current environment, it makes more sense to satisfy our cravings with ingeniously contrived substitutes like soy meat.

Sources

1. Khaitovich, P., et al. (2008). *Metabolic changes in schizophrenia and human brain evolution. Genome Biology, 9:* R124, 1-11.

2. Rudge, C. (1999). *Neanderthals, bandits and farmers: How agriculture really began. New Haven, CT: Yale University Press.*

3. Zaraska, M. (2016). *Meathooked: The history and science of our 2.5-million-year obsession with meat. New York: Basic.*

4. De Garine, I. (2004). *The trouble with meat: an ambiguous food. Igor de Garine, Hubert and R. Avila (Eds). Man and Meat. Estudios de l'Hombre, (19),* 33-54.

Is Religion Fiction?

If Jesus never lived, it would be another shocking example of religious fakery

As someone raised in a Christian country, I learned that there was a historical Jesus. Now historical analysis cannot find clear evidence that Jesus existed. If not, Christianity was fabricated, just like Mormonism and other religions. Why would people choose to believe?

Given the depth of religious tradition in Christian countries, where the "Christian era" calendar is based upon the life of Jesus, it would be astonishing if there was no evidence of a historical Jesus. After all, in an era when there were scores of messianic prophets, why would anyone go to the trouble of making one up?

In History, Jesus Was a No Show

Various historical scholars attempted to authenticate Jesus in the historical record, particularly in the work of Jesus-era writers. Michael Paulkovich revived this project as summarized in the Free Inquiry[1].

Paulkovich found an astonishing absence of evidence for the existence of Jesus in history. "Historian Flavius Josephus published his Jewish Wars circa 95 CE. He had lived in Japhia, one mile from Nazareth—yet Josephus seems unaware of both Nazareth and Jesus." He is at pains to discredit interpolations in this work that "made him appear to write of Jesus when he did not." Most religious historians take a more nuanced view agreeing that Christian scholars added their own pieces much later but maintaining that the historical reference to Jesus was present in the original. Yet, a fudged text is not compelling evidence for anything.

Paulkovich consulted no fewer than 126 historians (including Josephus) who lived in the period and ought to have been aware of Jesus if he had existed and performed the miracles that supposedly drew a great deal of popular attention. Of the 126 writers who should have written about Jesus, not a single one actually did so (if one accepts Paulkovich's view that the Jesus references in Josephus are interpolated).

Paulkovich concludes:

"When I consider those 126 writers, all of whom should have heard of Jesus but did not—and Paul and Marcion and Athenagoras and Matthew with a tetralogy of opposing Christs, the silence from Qumram and Nazareth and Bethlehem, conflicting Bible stories, and so many other mysteries and omissions—I must conclude that Christ is a mythical character."

He also considers striking similarities of Jesus to other God-sons such as Mithra, Sandan, Attis, and Horus. Christianity has its own imitator. Mormonism was heavily influenced by the Bible from which founder Joseph Smith borrowed liberally.

Mormonism Fabricated in Plain Sight

We may not know for sure what happened two millennia ago but Mormonism was fabricated in plain sight by a convicted conman. According to Christopher Hitchens[2]:

"In March 1826, a court in Bainbridge, New York, convicted a 21-year-old man of being a "disorderly person and an impostor." That ought to have been all we ever heard of Joseph Smith, who at trial admitted to defrauding citizens by organizing mad gold-digging expeditions and also to claiming to possess dark or 'necromantic' powers." Hitchens writes: *"Quite recent scholarship has exposed every single other Mormon "document" as at best a scrawny compromise and at worst a pitiful fake…"*

Smith's legacy was cleaned up via subsequent "divine revelations" that rejected first polygamy, and then racism, at convenient historical turning points. So, the historical development from fakery to respectable religion is well-documented.

There is no reason to believe that the genesis of any major religion was substantially different. This raises the question of why so many intelligent people choose to believe religious fictions. The most plausible explanation is that they cannot easily distinguish between organized religion and confidence rackets.

Starting a Fake Religion

Religious people may find that hard to swallow, so it is interesting to see what happens when someone sets out to found a fake religion. Would this work, or would members see through the deception?

American Indian film director Vikram Gandhi studied yogis and their followers in India. He concluded that these holy men were confidence tricksters, scores of whom plied their trade throughout India in the manner of the Jesus story.

The filmmaker wondered whether he could pass himself off as a guru here in the U.S. He cultivated a fake Indian accent, grew out his hair and beard and reinvented himself as Sri Kumare, a mystic hailing from a fictitious Indian village.

In the film, Kumare (2011), the director founds his cult in Arizona where he unloads his bogus mysticism upon the unsuspecting public. He soon draws a group of devoted followers who seek his counsel on their life problems and become frighteningly dependent upon his new-age advice.

The underlying psychology may be fairly simple. Common confidence tricksters work their magic by telling victims what they want to hear. The same is true of successful prophets who offer pie in the sky bye and bye[3]. The only reason that Jesus does not fit in this category is that he probably never existed.

Sources

1. Paulkovich, M. (2014). *God on Trial: The Fable of Christ. Free Inquiry. Vol 35, #5,*

2. Hitchens, C. (2007). *God is not great: How religion poisons everything. New York: Twelve.*

3. Barber, N. (2012). *Why atheism will replace religion: The triumph of earthly pleasures over pie in the sky. E-book, available at: Atheism-Will-Replace-Religion.*

Nigel Barber

Conservatives Big on Fear, Brain Study Finds
Are people born conservative?

Peering inside the brain with MRI scans, researchers at University College London found that self-described conservative students had a larger amygdala than liberals.

The amygdala is an almond-shaped structure deep in the brain that is active during states of fear and anxiety. Liberals had more gray matter, at least in the anterior cingulate cortex, a region of the brain that helps people cope with complexity.

The results are not that surprising as they fit in with conclusions from other studies. In 2010, researchers from Harvard and UCLA San Diego reported finding a "liberal" gene. This gene had a tiny effect, however, and worked only for adolescents having many friends. The results also mesh with psychological studies on conflict monitoring.

What It Means

There is a big unknown underlying these findings. Supposing that the size of one's amygdala really does increase the likelihood of being a conservative, is the size of the amygdala determined at birth, or does it perhaps increase with frightening childhood experiences, such as authoritarian parenting and corporal punishment?

Similarly, one might ask whether the gray matter difference is affected by exposure to educational challenge, social diversity, or childhood cognitive enrichment.

The "born" versus "acquired" perspective on political attitudes is important to psychologists. After all, if political proclivities are fixed at birth in terms of brain anatomy, there is little hope of change. Most of us would probably like to see a world in which political attitudes were less polarized and more changeable, but that may be a pipe dream.

Meanwhile, the neuro-scientific fact of two very different political creatures helps clarify much of the political antics of modern democracies.

Most societies are divided into a party that wants change (the more liberal party) and one that is afraid of change (the conservatives). The liberal party is generally more intellectual and the conservative party is more anti-intellectual.

The conservative party is big on national defense and magnifies our perception of threat, whether of foreign aggressors, immigrants, terrorists, or invading ideologies like Communism. To a conservative, the world really is a frightening place.

Given that their brains are so different, it is hardly surprising that liberals and conservatives should spend so much time talking across each other and never achieving real dialogue or consensus.

If everyone was born with the same brain potential to acquire either conservative or liberal views, then we could be more optimistic about prospects for political communication and consensus-building. If voters were of like brain, perhaps they could be of like mind.

Source:
https://www.ucl.ac.uk/news/2010/dec/left-wing-or-right-wing-its-written-brain

Why Are Humans and Dogs So Good at Living Together?

Do dogs or humans gain more from our ancient association?

D ogs have a special chemistry with humans that goes back many tens of thousands of years. Researchers investigated this special evolutionary relationship from a number of different angles. Their results are surprising.

The social unit

Domestic dogs are descended from wolves so recently that they remain wolves in all biological essentials, including their social behavior. Wolf packs have some intriguing parallels with human families:

- They are territorial.
- They hunt cooperatively.
- Pack members are emotionally bonded and greet each other enthusiastically after they have been separated.
- In a wolf pack, only the alpha male and female are sexually active even though other pack members are sexually mature.

The social adaptations of dogs and humans are similar enough that dogs can live perfectly happy lives surrounded by humans, and vice versa. Dogs are pampered with the best of food and medical care, frequently sleeping in their owners' comfortable beds.

A family member

Why do people lavish so much care on a member of an alien species? A short answer is that on an emotional plane, families do not see the dog as alien. According to John Archer[1] of the University of Central Lancashire, who has conducted a detailed study of dog-human relations from an evolutionary perspective, about 40% of owners identify their dog as a family member reflecting social compatibility between our two species.

Dogs are extraordinarily attentive and have an uncanny ability to predict what their owners will do, whether getting the dog a meal or preparing to go on a walk. Experiments show that dogs and wolves can be astute readers of human body language, using the direction of our gaze to locate hidden food[2] - a problem that is beyond chimps.

Dogs also seem attuned to the emotional state of their masters and express contrition when the owner is annoyed, for example. The capacity to express affection, unconditionally, makes the dog a valued "family member."

Domesticating each other?

Dogs were the first domestic animal with whom we developed a close association. Mitochondrial DNA research suggests that most domestic dogs have been genetically separate from wolves for at least 100,000 years. This means we have associated with dogs for as long as we have been around as a species (Homo sapiens). Indeed, some enthusiasts, including Colin Groves of the Australian National University, in Canberra, believe that our success as a species is partly due to help from dogs[3].

According to Groves: *"The human-dog relationship amounts to a very long-lasting symbiosis. Dogs acted as human's alarm systems, trackers, and hunting aides, garbage disposal facilities, hot water bottles, and children's guardians and playmates. Humans provided dogs with food and security. The relationship was stable over 100,000 years or so and intensified in the Holocene into mutual domestication. Humans domesticated dogs and dogs domesticated humans."*

Relying on dogs to hear the approach of danger and to sniff out the scent of prey animals, our ancestors experienced a decline in these sensory abilities compared with other primates. This conclusion is confirmed by shrinkage of brain regions devoted to these senses (the olfactory bulb and lateral geniculate body).

During the long period of our association, dogs' brains have shrunk by about 20 percent, typical for animals such as sheep and pigs who enjoy our protection. Domesticated animals undergo tissue loss in the cerebral hemispheres critical for learning and cognition. If we relied on dogs to do the hearing and smelling, they evidently relied on us to do some of their thinking. If Groves is correct that dogs have domesticated humans, then the human brain would also have gotten smaller. Surprisingly, human brains have actually shrunk, but by only a tenth, suggesting that dogs got more out of the deal than we did.

Sources

1. Archer, J. (1997). *Why do people love their pets? Evolution and Human Behavior, 18, 237-259.*

2. Udell, M. A. R., Dorey, N. R., & Wynne, C. D. L. (2008). *Wolves outperform dogs in following human social cues. Animal Behaviour, 76, 1767-1773.*

3. Groves, C. P. (1999). *The advantages and disadvantages of being domesticated. Perspectives in Human Biology, 4, 1-12.*

Top 5 Signs That Women are Converging with Men
Modern women take greater risks and are more interested in casual sex

Fairly rigid gender divisions of the past are giving way to a much more equal relationship. Women are beginning to act and feel more like men even as men's actions and sensibilities are converging with women. What are the unmistakable signs that women are becoming more like men?

1. Participation in Workforce

In the evolutionary past, women likely contributed more food from gathering than men contributed from hunting. In agricultural societies, and in post-industrial-revolution ones, men adopted the breadwinner role, and married women were more dependent on their husbands.

This trend is reversing as more married women participate in the paid workforce. In 1900, about half of single women worked for pay but most stopped working at, or soon after, marriage, and never worked again. Labor participation rates of married women in the U.S. soared from 6 percent in 1900 to 61 percent in 2000[1]. This implies that their economic role is equivalent to that of men.

Women are also much more ambitious, as illustrated by their rising share of bachelor's degrees, from a small minority of 19 percent in 1900 to a distinct majority of 56 percent in 2000[1]. Female competitiveness in jobs and education is underscored by increased interest in sports.

2. Increased Participation in Contact Sports

Female participation in sports is historically much lower than male participation. From an anthropological perspective, sports are associated with warfare, both as a form of physical training, and also as a way of demonstrating physical preparedness for battle. Neglect of female sport reflected a belief that physical competition (like warfare) was more masculine than feminine.

85

Low participation rates by women in sports was partly due to a lack of facilities that 1972 Title IX legislation sought to remedy. At this time, there were only 4 percent of U.S. high school women participating in team sports. Participation shot up to 25 percent within a quarter century bringing female participation close to that of males.

3. Increased Risk-Taking

Modern women are behaving much more like men when it comes to risk-taking and aggression. One sign of this phenomenon is greater participation in contact sports and dangerous competitions such as horse racing or car racing. According to Anthropologist Elizabeth Cashdan[2], in societies where women compete more amongst each other whether in occupations, or over spouses, their level of testosterone increases. High testosterone is correlated with risk-taking and aggression.

There are far more women driving on the roads today and they drive more aggressively and dangerously than ever before. Consequently, their accident rates have risen from very low levels. Young women are almost as dangerous on the roads as young men whose aggression and recklessness make driving so much more dangerous for everyone else. Reckless driving is associated with increased use of alcohol and recreational drugs.

4. Increased Physical Aggression

Physical strength is clearly one risk factor for committing violent crimes which helps explain why so many of the perpetrators are men. Another key reason for the gender difference is that men fight over women more than women fight over men.

In the past, female involvement in organized crime was minimal and largely due to association with gangster husbands or boyfriends. All that is changing with women beginning to claim a slice of the action as gender equality moves into violent crime as well as other high-risk activities.

As women have begun to take leadership positions in large corporations, they have also acted as leaders in criminal enterprises. One of the most successful Latin American drug kingpins was a Colombian woman, Griselda Bianco, known as La Madrina, who ran an extensive U.S. operation from Miami.

Given the many other ways in which women have come to resemble men of the past, it is no surprise to learn that their level of violent crime is on the rise. This is showing up both as fighting between girls of school age and participation in more serious crimes, such as aggravated assaults where female rates are soaring. The proportion of women in the US correctional system doubled between 1985 and 1998 (from about 0.5 to 1 percent, 3)

5. Increased Interest in Casual Sex

Whether they are violent or not, women might reasonably be much less interested in casual sex than men based on the greater biological investment by women than men in children. Since males give less than they get in terms of biological investment in offspring, they are more eager to mate. As countries become more affluent, however, women get more interested in casual sex and masculine interest declines so that the gender difference vanishes[4].

Why do women in developed countries become more interested in casual sex? One reason is that contraception removes most of the risk of unwanted pregnancy. Another factor is that their hormone profile changes in response to competition at work, and competition over mates, thereby increasing sex drive and risk taking[2]. Moreover, in an environment where three-quarters of women are sexually active before marriage by age 19[1], the coy strategy of postponing sex until after marriage is not a winner.

Sources

1. Caplow, T., Hicks, L., & Wattenberg, B. J. (2001). The first measured century: An illustrated guide to trends in America, 1900-2000. La Vergne: TX: AEI Press.

2. Cashdan, E. (2008). Waist-to-hip ratios across cultures: Trade-offs between androgen- and estrogen-dependent traits. Current Anthropology, 49, 1099-1107.

3. Greenfeld, L. A. and Snell, T. L. (1999). Women offenders. Washington, DC: U. S. Department of Justice Bureau of Justice Statistics.

Is the Modern World More Violent?

Or is it journalism that is red in tooth and claw?

If it bleeds, it leads is a truism of news coverage. We all sympathize with the victims of senseless violence, and their families, because we know that it could have been us, and our families. Yet, our world has never been less violent – except in news media and entertainment.

This journalistic bias has two adverse effects. First, it makes news junkies worry unduly. Second, it encourages rampage killers by giving them instant "celebrity."

Media coverage exploits our sympathy and empathy for victims, elevating our sense of danger out of proportion to the actual threat. Of course, it also feeds on the shock of seemingly safe places being violated, such as churches.

Belief in a Violent World

While the world is a lot less violent today than at any other time in history, or prehistory, that fact escapes us thanks to our daily diet of journalistic carnage. The worldwide probability of dying in a terrorist attack is infinitesimal at less than one in a million per year[1]. This risk is about three times lower than it was in the 1980's. Yet, respondents to surveys believe that the risk has gone up[2] - a phenomenon that can be attributed to extensive coverage of spectacular terrorist attacks, such as the Al Qaeda attacks of 9/11, 2001.

A similar point can be made about fear of violent crime. Homicide rates in Western Europe today are only about a fortieth of what they were in the 14th century[1]. Homicide rates in the US are about four times higher than in Europe, but they are substantially lower than during the colonial period. Moreover, homicide rates today are only around half what they were in 1990, a point that is easily missed if one likes to watch the TV news.

Exaggerated fear of crime and terrorism can have very damaging consequences for our health and well- being. If we fear flying to some destination because of terrorism and choose to drive instead, our risk of dying is greatly increased because flying is much safer than driving[2]. Similarly, exaggerated fear of violent crime can inhibit walking and exercise, with very damaging health consequences. Of course, such fears contribute to clinical anxiety and depression. More balanced journalism can thus save lives and contribute to happiness. As it is, the "if-it-bleeds-it-leads" approach may actually contribute to the worst acts of violence.

Reinforcing Appalling Violence

One of the biggest problems about disproportionate coverage of wanton violence is that many homicidal maniacs crave publicity. This fact emerged in relation to the Zodiac killings of the 1960s and 1970s where a serial killer played sardonic cat-and-mouse games with the authorities via messages published in newspapers.

Media today are too sophisticated to fall for such blatant manipulation, but the truth is that doing something really awful remains a guaranteed method of achieving instant fame, or infamy (a distinction that often seems paper thin).

The worst rampage killers get several days of concentrated media attention - a glare so bright that marketing experts in huge companies, such as American Airlines cannot resist getting in on the act (by offering a scholarship in the name of the slaughtered pastor). At the end of their sorry lives, the murderers can reflect that their crime was so big it took the President of the United States to comfort the victims.

Sources

1. Pinker, S. (2011). *The better angels of our nature: Why violence has declined.* New York: Viking Penguin.
2. Mueller, J. (2006). *Overblown: How politicians and the terrorism industry inflate national security threats and why we believe them.* New York: Free Press.

Why People Conform

Despite costs to individual liberty, we comply with whims of groups

Social psychologists found people highly susceptible to social influences as demonstrated by classic experiments in obedience and conformity. Why might we be so malleable? One intriguing clue is offered by subsistence ecology, specifically the transition to agriculture.

The Ecology of Conformity

Personal autonomy was very important to humans in hunter-gatherer societies[1] whereas agricultural societies emphasized obedience to authority figures.

The individuality of hunter-gatherers may be illustrated by the willingness of women to engage in many extramarital relationships despite the fact that these sometimes brought them a great deal of trouble[2].

Hunter-gatherer defense of egalitarian principles is reflected in their political structure which is exceptionally flat. Each forager group typically has a single headman, or head woman, who has few perks of office and spends a lot of time resolving other people's interpersonal disputes.

Male hunters are highly individualistic because they must often rely upon their own skill and endurance in bringing down large game animals, even though hunting is a group activity[3]

Matters are very different in subsistence agriculture where success in farming is usually determined by adherence to time-honored practices of cultivation and crop rotation. Children must also stick by their parents for many years if they wish to acquire land ownership or farming rights. For this reason, agricultural societies emphasize respect for elders and obedience to time-honored traditions.

It is no accident that religious zealotry is most likely to be found in agricultural societies where the individual is raised to accept the views and beliefs of the community without question. Such conformist tendencies may facilitate inculcation in close-knit groups from warrior castes to religious communes.

Costly Groups Like Communes and Warrior Castes

Members of some groups conform to very exacting demands. Some religious sects expect their members to give up all their property upon joining. Others prescribe inconvenient dress styles, as in the case of Hassidic Jews who sweat out the summer wearing clothes that would be more suited to a Russian winter.

Why do members concede to so many apparently unreasonable requirements? One clue is provided by research on historical communes. Those having more requirements of members were found to last longer. This suggests that the high price of admittance increased group cohesion[4].

Other evidence shows that the more people are required to invest in a group, the more connected to that group they feel, and the more likely they are to behave altruistically towards fellow members. For example, Muslims who attend worship regularly are more likely to approve of religious martyrdom[5].

A similar logic extends to many other groups from book clubs to warrior castes. Warriors sometime undergo grueling initiation ceremonies that follow stringent rules of conduct involving painful tests such as scarification or ritual circumcision. Warrior groups having the largest costs of membership (conformity to a string of arbitrary rules and privations) also identify most strongly with their warrior group[6].

Behavioral conformity is often possible without any depth or sincerity. Many religions impose the further requirement of conforming to a detailed belief system that often bears a tenuous connection with reality. For example, Christians expect their members to accept that the communion wine turns into blood prior to being consumed. Why do rational people go along with beliefs that defy what we know about the natural world?

Religious Conformity as a Contract

Perhaps believers believe what they do because they are obliged to do so. In some Islamic countries, disbelief is criminalized as a capital offense, even if the death sentence is rarely carried out. Similarly, changing one's religion is theoretically punishable by death. Such customs seem draconian in a pluralistic society but are less objected to in a world of religious homogeneity. In earlier times, Christian "heretics" were burned at the stake for seemingly minor theological divergences.

Conformity of belief is analogous to conforming with ritual obligations of a religion. In other words, it is another cost of membership. Just as members agree to prescribed customs and ritual obligations, they also sign on to an entire system of beliefs.

It is as though they had entered a contract according to which they implicitly agreed to believe what the group in general believes, however far-fetched, implausible, or scientifically dubious. By this reasoning, belief in what seems highly improbable to outsiders is a test of commitment to the religion. If conformity to a religion has disturbingly irrational elements, the same issue is often raised in connection with political partisanship.

Political Conformity

Conformity to the beliefs of political parties has many parallels with the acceptance of religious beliefs. In some cases, political affiliation affects how people describe themselves in terms of how religious they are, with conservatives emphasizing their own religiosity. They also emphasize family ties and tradition, and place great importance on social conformity and obedience to authority[7,8].

In the past, I have suggested that these differences are predicated on varied approaches to securing investment by older generations in children. There are several reasons that being close to family members might favor reproductive success. One is that senior family members may control resources, such as land, or property, that go preferentially to children who stay close by.

If conservatives are more likely to conform to social expectations and to focus on their in-groups, liberals are more open to social diversity and new experiences. This means that they are attracted to a more bohemian way of life and more unconventional career paths, such as being artists and writers.

Artists and Eccentrics

Artists share with eccentrics a willingness to break the mold, behaving differently from earlier generations and other people more generally. They may do so because they are more socially detached. This is frequently due to experiencing diversity early in life, whether as a result of immigration, gender identity, illness, or other influences that make them see the world as substantially different from the mainstream. This is an important key to creativity. We all benefit when they escape the pressure to conform.

Sources

1. Boehm, C. (2000). *Hierarchy in the forest.* Cambridge, MA; Harvard University Press.
2. Shostak, M. (1981). *Nisa: The life and words of a !Kung woman.* Cambridge, MA: Harvard University Press.
3. Berry, J. W. (1967). Independence and conformity in subsistence-level societies. *Journal of Personality and Social Psychology, 7, 415-418.*
4. Sosis, R., and Bressler, E. R. (2003). Cooperation and commune longevity: A test of the costly signaling theory of religion. *Cross- Cultural Research, 37, 211-239.*

The Blank Slate Controversy

How much of our individuality is determined at conception?

Psychologists such as B. F. Skinner used to argue that people were blank slates in the sense that almost all our behavior was learned. Evolutionary psychologists begged to differ. Who is correct?

The blank slate idea has a long history in philosophy that goes back to Aristotle. Skinner's version draws on English philosopher John Locke who developed a theory of knowledge as formed by the association of sensory experiences (and referred to a blank sheet of paper).

Brain Development is Hard to Predict

One way of deciding between the blank slate and biological determinism is to investigate the impact of unusual events in prenatal development.

Various lines of evidence indicate that if female fetuses are exposed to excessive levels of sex hormones, they grow up with masculinized behavior[1]. Amongst experimental monkeys, females exhibit the male pattern of more vigorous physical activity. For humans, women who were exposed to high levels of sex hormones in the womb - because their mothers mistakenly continued to take birth control pills - more self identified as lesbians.

In the case of androgen insensitivity, a genetic disorder, persons having a male genotype can grow up looking, and behaving exactly like women. This natural experiment indicates that exposure to androgens during development determines much of masculine appearance and behavior.

Brain function is strongly affected by exposure to stress hormones so that children growing up in stressful homes where parents fight a lot score lower on IQ tests[2].

The impact of prenatal nutrition on intelligence is well established. Researchers are finding that well-nourished mothers give birth to children who grow up not just to be taller and healthier but also more intelligent, more motivated to work hard, and more economically successful[3].

Critics of such evidence might argue that it involves something going awry in development that is of questionable relevance for people growing up under more normal circumstances. Even so, it is reasonable to assume that there is a range of variation for nutrition, stress hormones, and sex hormones within which normal development may occur. So, at the very least, we can conclude that who we are is very much affected by biological factors within the womb, contrary to blank-slate theories.

A different approach is to begin from birth and ask whether many aspects of our personality and cognitive function are already decided by our genetic heritage and biology.

What Behavior is "Loaded" at Birth

Perhaps the clearest evidence against the blank slate concept is the fact that people remain much the same throughout their lives on personality dimensions. Some of us are extroverts. Others are introverts. Some of us are physically very active whereas others are less energetic. Some of us are highly emotional in response to minor events in our lives whereas others are unperturbed.

Such personality traits are predisposed by the biology of our brains and firm evidence in favor of this view comes from the fact that these traits are strongly heritable (with genetic ancestry accounting for around half of individual differences in major personality dimensions[4].

How our brains process information is, to some degree, predetermined by brain anatomy and physiology. Neuroscientists have developed a detailed knowledge of the functional anatomy of the brain so that damage to a particular part yields predictable functional deficits. Damage to the hippocampus causes memory problems, for example.

Moreover, information travels in predictable paths within the brain. So visual information travels from the retina to the thalamus but must reach the cortex for complex pattern recognition to be accomplished and visual perception to occur.

Despite the evidence for behavioral predispositions being present at birth, the brain itself has blank-slate-like properties[5]. This phenomenon was studied most extensively for cortical cells that initially do not know what they are supposed to do but develop a rapport with neighboring cells, responding most strongly to those that stimulate them the most. This means that if a person were to lose a finger, the parts of the cortex that had represented that finger will likely start responding to input from another finger.

The inherent changeableness, or plasticity, of brain cells has been compared to a blank slate upon which sensory and motor input gets written. Sometimes, that information may get erased so that something different may be written in its place.

So, Who is Right?

The historical fault lines in psychology lie between the blank-slate model of the behaviorists and the opposite extreme favored by many evolutionary psychologists.

Both extremes would appear to be wrong. We know that personality traits are strongly influenced by genotype, for example, contrary to the behaviorist perspective. Some plausible mechanisms through which this occurs have been described, such as genetic modification of the number of neurotransmitter receptors.

On the other hand, the more we discover about the brain, the more we are impressed by its capacity to respond to changing sensory inputs. So, it is quite implausible that people would be born with a fully functional genetically determined brain "program" that solved some Darwinian problems such as preventing spousal infidelity.

The brain may not be entirely blank at birth, but it is not entirely programmed either. It is an interesting mix of script and improvisation.

Sources

1. Barber, N. (2002). *The science of romance.* Buffalo, NY: Prometheus.

2. Delaney-Black, V., Covington, C., Ondersma, S. J., Nordstrom-Klee, B., Templin, T., Ager, L., et al. (2002). *Violence exposure, trauma, and IQ and/or reading deficits among urban children.* Archives of Pediatric and Adolescent Medicine, 156, 280-285.

3. Case, A. & Paxon, C. (2008). *Stature and status: Height, ability and labour market outcomes.* Journal of Political Economy, 116, 491-532.

4. Plomin, R. (1990). *Nature and nurture: An introduction to human behavioral genetics.* Belmont, CA: Wadsworth.

5. Kalisman, N., Silberberg, G, and Markram, H. (2005). *The neocortical microcircuit as a tabula rasa.* Proceedings of the National Academy of Sciences, 102, 880-885.

Why We Make Love at Night
The reasons range from practical to romantic

Most people make love when they go to bed, which is usually at night[1]. "Sleeping with" someone has thus become a synonym for having sex. Why human sex and sleep are so intertwined, though, remains mysterious. There is no obvious reason why copulation under the covers at night is biologically optimal. Anthropologists note that sexual infidelity can occur opportunistically at any time of day, beginning with the dawn. That is when hunter-gatherers typically leave their huts to urinate. It is also when couples can have quick trysts out of earshot of their sleeping spouses. But while such meetings may accomplish the biological job of fertilization, they leave much to be desired from a variety of perspectives—physiological, social, psychological, and ethical.

Why Sex Before Sleep May Be Preferable

Researchers find that most marital sex occurs around bedtime. More than half of sexual encounters occur between the hours of 10:00 p.m. and 2:00 a.m. with a smaller additional peak at 6:00 a.m. when couples are likely to be waking up[2]. Couples are more likely to have sex on weekend nights, suggesting that work schedules dictate patterns of sexual activity to some extent. Avoiding sex on work nights may help employees feel better rested the following day.

The strong circadian pattern of sexual activity has a fairly simple explanation in terms of availability or opportunity. Married people are more likely to make love at the time they go to bed because they are available to each other. Of course, that begs the question of why married couples generally sleep together in the first place. Even if this question is ignored, one is left with the problem of why sex is more common in the evening than in the morning.

Physiological Explanations

For most mammals, copulation is brief, although there are exceptions, such as in dogs, where mates remain joined due to coital lock that may serve to foil male competitors. The brevity of mating does not interfere with fertilization, however. The same may not be true of upright walkers like humans, for whom gravity would tend to remove ejaculate from the reproductive tract. If so, lying down after sex may increase the chances of conception. Lying down together may also contribute to intimacy and sexual pleasure, which have been thought to affect conception as well.

Relationship Explanations

A couple sleeping together has diverse implications for pair bonding and relationship strength. Of course, there are many possible reasons for sleeping together that have little to do with intimacy. A couple is more effective at conserving heat during the cold of night or in winter, which has survival implications for indigenous peoples like the Inuit. Similarly, two pairs of ears are better than one when it comes to detecting nighttime risks, such as attacks by wild animals or enemies.

On the face of it, though, lying down in close physical contact promotes intimacy by increasing oxytocin production. This is the "cuddling hormone" that promotes relationship strength for many mammals, including powerful mother-infant bonds[3]. Spending time in close proximity thus contributes to the strength of pair bonds—not just in humans but in many other mammals, including the humble prairie vole where this has been extensively studied[4].

For pair-bonded species, a close relationship is critical for success in raising offspring. Indeed, females prefer to mate with their partner and may refuse to mate with unfamiliar males.

This pattern applies to humans who satisfy most of the criteria for being a pair-bonded species[3]. In that context, it is understandable that sexual relations occur most often during bedtime hours if prolonged physical contact promotes feelings of closeness and intimacy.

Sources

1. Palmer, J. D., Udry, J. R., and Morris, N. M. (1982). *Diurnal and weekly, but no rhythms in human copulation. Human Biology, 54, 111-121.*

2. Refinetti, R. (2005). *Time for sex: Nycthemeral distribution of human sexual behavior. Journal of Circadian Rhythms,* DOI: 10.1186/1740-3391-3-4

3. Barber, N. (2002). *The Science of Romance. Buffalo, NY: Prometheus.*

4. Insel, R., and Hulihan, T. (1995). *A gender-specific mechanism for pair bonding: Oxytocin and partner preference formation in monogamous voles. Behavioral Neuroscience, 109, 782-789.*

Why We Consume So Much

Affluent societies have too much but want even more

In agricultural societies of the past, farmers were mostly self sufficient. They worked hard but consumed little. Today we consume so much that the oceans are filled with trash. What changed?

Rising Disposable Income

A great deal has happened. To begin with, rising global affluence following the Industrial Revolution put a lot more money in people's wage packets[1]. The proportion of pay being spent on non-essentials like food and shelter declined, leaving a surplus of "disposable income".

This had several important consequences. Social mobility took off. Workers began using their disposable income to communicate social success and status.

This could be one reason that the bulk of disposable income is devoted to leisure activities. Of course, such activities are also pleasurable in themselves, from enjoying sunshine in a vacation spot to fine dining or driving luxury vehicles.

In feudal societies there had been minimal social mobility and a luxurious lifestyle was the exclusive preserve of the landed elite.

This scenario changed profoundly with the rise of the middle classes following the Industrial Revolution. Some scholars argue that the lives of ordinary people in developed countries today are more luxurious in material terms than those of the most sybaritic monarchs of the past (such as Louis 16th of France) given the variety of consumer goods available[2].

The Rise of Consumerism

Social mobility preceded the steam-engine-based Industrial Revolution. Cloth manufacturers "put out" their orders, or subcontracted them, to cottage weavers. Using tallow candles that shed a bright and steady light, cotters worked night and day to fill their orders and saw a modest increase in their standard of living[3].

Due to rising wages, a person could save to buy a better home, or better furniture, or more elegant clothing. With social mobility, every person was responsible for their own station in life. Keeping up with the Joneses was born.

The period preceding the factory production of cloth is aptly named the "Industrious Revolution[3]." It was accompanied by slow but steady economic growth culminating in the modern era of cheap global shipping and runaway consumption.

When living standards rise, beneficiaries are never really content with where they are. There is always someone else who has a better home, a better car, better furniture, or sends their children to better schools, or spends more on cosmetic surgery.

Aspiring to lead a more luxurious life is a treadmill that keeps consumers in debt and constantly working harder to get where everyone else seems to be. This is a positive climate for businesses.

Media, Advertising, and Artificial Needs

The desire for more goods and services is enabled by modern communication technologies. This played out following the introduction of television to towns in the US West. The rise of petty thefts that followed was motivated by dissatisfaction of viewers with their standard of living compared to that depicted on the small screen[4]. In marketing terms, demand for many consumer goods was created overnight.

The same phenomenon occurs on the Internet where users are exposed to luxurious living by celebrities around the globe. Of course, Internet shopping is the ultimate enabler for luxury goods because they are rarely more than a few clicks away.

The desire for luxurious, and status-signaling, products is a powerful motive for workers. Indeed, Thorstein Veblen, an early sociologist, argued that "conspicuous consumption" is the main reason that people with money choose to spend it[5].

The carrots of consumption may well be more potent than the sticks of management in boosting work effort. Ordinary consumers today behave rather like the elites of the past with the key difference that elites do not need to work. Even the poorest segment of our society gives in to cravings for luxury goods from designer watches, shoes, and sunglasses to electronic devices and cars.

Modern societies are something of a rat race where people work hard at consuming during their off time and work even harder during their hours on the job so that they can earn enough to pay their bills. The irony is that although we live like the kings of history in terms of material standard of living, we do not recognize this reality[1,2].

The flavor of this discontent is captured in the witticism that a rising tide lifts all yachts (rather than all boats). Improvements in living conditions in developed countries are real, substantial, and historically unprecedented, although pessimists find it hard to accept and focus upon the unequal distribution of wealth.

Life expectancy of US residents doubled in the 20th century, for example, but this fact is lost today in concern over the fact that rich people live longer.

We would also like their money. Avarice is a cruel master. Despite having too much, we are tormented by the desire for more, knowing that other people are a lot better off than we are.

Sources

1. Floud, R., Fogel, R. W., Harris, B., & Hong, S. C. (2011). The changing body: Health, nutrition, and human development in the Western world since 1700. Cambridge, England: NBER/Cambridge University Press.

2. Ridley, M. (2010). The rational optimist. New York: Harper Collins.

3. Galor, O., and Moav, O. (2000). Natural selection and the origin of economic growth. Quarterly Journal of Economics, 117, 1133-1191.

The Autism-Genius Connection

Chess champion Magnus Carlsen illustrates social costs of unusual creativity

S ome people are exceptionally good at focusing on rarefied abstract problems. Some of these have exceptionally high IQ. Some are autistic. Some are both.

This link was raised in a recent movie (Magnus) chronicling the life of world chess champion Magnus Carlsen, who emphatically denies that he is autistic. This denial is treated skeptically in the movie, however.

Tuning in to a Talent, Versus Tuning Out Other People

One of the key paradoxes in the lives of highly creative people is that high achievement is promoted by an exclusive focus on some one field of endeavor. Whether it is Edison inventing the electric light bulb, or Beethoven writing a symphony, the capacity to transcend prodigious challenges requires a keen ability to screen out distractions, whether social, or practical. Highly creative people are not inherently asocial, of course, and neither are autistic individuals, whatever problems they might have in negotiating the social landscape.

Even so, tuning in on a talent often requires tuning out other people. In the documentary movie, Magnus, the protagonist spends much of his time absorbed in the world of electronic chess where family members get tuned out. This illustrates the social costs of unusual creativity in any field.

Chess is a social game, of course, and it involves a fair amount of gamesmanship like showing up at the last minute to unsettle an opponent. As in boxing, the confidence game—getting inside an opponent's head—is often as important as the player's actual moves.

Chess experts distinguish opening moves that can be memorized to some extent, from the middle game that is a great deal more fluid, or improvisational. Carlsen uses a great variety of opening moves, making it

106

difficult for opponents to prepare. He excels in the middle game and overwhelms opponents with creative moves and counter moves. At that point, it is as though an inexperienced sword fighter finds themselves in a duel to the death with a tireless opponent who has buried all previous challengers.

Autistic Savants

Autistic savants are people who suffer from a developmental disability yet demonstrate some cognitive ability, exceeding what most people can do[1]. These abilities may involve art, music, arithmetic, spatial skills, or calendar abilities where the savant can determine what day of the week some date is with speed and accuracy. About half of savants are autistic, the others suffer from some form of brain injury or disease. About a tenth of autistic people have savant abilities.

Another common skill manifested by savants is prodigious feats of memory. As a small child, Carlsen amused himself by memorizing the capital cities, and flags, of all the countries of the world. Such narrow and intense focus is characteristic of autism, as is his ongoing preoccupation with Donald Duck movies.

Individuals with autism spectrum condition often exhibit unusual skills in attention and perception relative to the general public[2]. This might be helpful in memorizing chess moves. In fact, this capacity—that is uniquely developed in autistic individuals—may be a key to developing unusual talents.

Autism and Talent

Exceptional talent in autistic individuals may begin with sensory hypersensitivity that makes many uncomfortable in proximity to loud noises, for example[2]. They have great attention to detail and organize the perceptual world to an unusual degree—a phenomenon referred to as hyper-systematizing.

This propensity even affects observable behavior where daily routines are repeated in exactly the same way. Even small children express a craving for organization by arranging their toys in neat lines.

Hyper-systematizing propensities may be expressed in savant-like activities, such as calendar abilities that are not very socially useful, or they may contribute to socially celebrated achievements in the visual arts, or chess. Whether Carlsen is on the autistic spectrum or not, his exceptional chess skills at least bear analogy with the accomplishments of autistic savants.

Adaptive Significance

At present, the brain basis of autism is poorly understood and the same is even more true of savant abilities. One of the great mysteries of brain development is the way that even as some capacities are sharpened - as a youngster acquires efficient contact with the social and non-social environment - others are tuned down.

This phenomenon is illustrated by the savant abilities of attention, memory, and the autistic trait of hyper-systematization which are far less developed in people outside the autistic spectrum. This speaks to compromises and limitations in brain development and function.

The same general principle applies to adult brain function. For example, the decay of some old memories may improve brain function by getting rid of clutter. Similarly, inattention to details that are perceived as trivial may be an inevitable side effect of having a brain that sorts events according to their emotional significance so that beautiful strangers walking down the street are more likely to register than their less attractive counterparts.

The great range of human cognitive capacities suggests that there is no single optimum when it comes to brain development, or function. Some people function well despite having limited capacity for memorization whereas others have almost unlimited recall. Some are

exceptionally good at attending to details whereas others quickly
get bored.

Given that boredom is a leading problem in the workforce, employers
are discovering that autistic people make ideal employees in some jobs,
including human resources, benefits administration, and coding where
attention to detail is very important. This is an everyday example of the
specialized autistic abilities that contribute to more glamorous
achievements in art, chess, and music.

Source
*1. American Psychiatric Association (2013). Diagnostic and statistical manual of
mental disorders (5th ed.). Arlington, VA: American Psychiatric Publishing.*
*2. Baron-Cohen, S., et al., (2009). Talent in autism: Hyper-systematizing, hyper-
attention to detail and sensory hyper-sensitivity. Philosophical Transactions of the
Royal Society of London B Biological Sciences, 364, 1377-1383.*

Gender Fluidity and Hormone Disruptors
Hormone-disrupting chemicals may increase gender dysphoria

Polluted groundwater yields ambiguous genitalia for vulnerable species. The modern human environment is replete with substances that mimic sex hormones. Could these chemicals play a role in contemporary gender fluidity?

Hormone Disruptors

The Age of Chemistry solved all kinds of problems, from flexible new materials to more productive agriculture. The benefits were obtained at the cost of many insidious problems. These include damage to critical ecosystems and a variety of toxicity-related diseases and developmental disorders.

Such issues were highlighted by research showing that polluted ponds could turn male frogs into females. Interestingly, some of the affected frogs were capable of reproducing, but produced all male offspring that had the effect of crashing the population.

Such problems are not restricted to wildlife occupying heavily polluted areas. They are also found in humans leading their lives in apparently clean homes and non-toxic environments.

One casualty of under-the-radar hormone disruptors is gender development.

Hormone Disruptors and Gender Development

Humans may be vulnerable to hormone disruptors in ways that resemble the effects on frogs inhabiting polluted wetlands.

The culprits are products, such as cosmetics, that are not ordinarily considered dangerous, or polluting. One example concerns phthalates that are present in plastic packaging.

Another candidate is pharmaceuticals. This phenomenon emerged in connection with the contraceptive drug diethylstilbesterol. Mothers who were unaware of their pregnancy continued to take the drug. It was found to have masculinizing effects on the brains of female fetuses. The limited evidence available indicated that when they matured, such females were less interested in caring for children and more interested in female romantic partners.

Recent evidence indicates that phthalates from plastic and polychlorinated biphenyls (PCBs) are one of many factors predicting gender dysphoria, particularly in the case of male-to-female transgenders.

Chemistry Versus Psychological Openness
Of course, the LGBTQ community rejects the chemical reduction of what is seen as a voluntary choice and is hostile to the notion that a non-binary orientation is somehow the result of a disorder.

Whatever one's opinions about such matters, it is short-sighted to ignore actual environmental threats to brain development.

Gender dysphoria is certainly not reducible to any single cause, chemical or otherwise. Indeed, there are several psychological correlates. These include any kind of trauma experienced during childhood. Another possible factor is abusive parenting.

Real as these psychological pathways may be, they do not invalidate, or even weaken, the hormone disruptor theory. Indeed, they suggest a similar nexus in brain development. Stress hormones also affect gene expression and have far-reaching consequences for subsequent psychological development in addition to gender identity.

The role of industrial toxins in disrupting human brain development surfaces in reduced IQ scores for children raised in lead-polluted inner cities. Other heavy metal pollutants produce similar consequences.

The Age of Chemistry Versus Health and Reproduction

Researchers are beginning to recognize – based on limited available evidence – that the chemical environment is one of many available explanations for gender dysphoria in modern environments.

On the one hand, this phenomenon is a canary in the coalmine offering yet one more clue that we are endangering our health and well-being through incautious acceptance of modern materials that pose endocrine threats.

The semblance of comfort and convenience that products such as plastic wrapping and cosmetics present is the sheep's clothing beneath which major threats to our species may reside. For, if hormone disruptors affect gender orientation and gender dysphoria, we can be sure that they are a clear and present danger to reproduction and the long-term viability of our species.

If so, they join a long list of ways that our industrial society is doing us in.

Sources
https://news.berkeley.edu/2010/03/01/frogs/ male frogs turned into
    females (that may reproduce and produce all males)
https://www.ncbi.nlm.nih.gov/pubmed/24793993 pthalates and
    polychlorinated iphenyls (PCBs) impair development of fetal testes
https://www.ncbi.nlm.nih.gov/pmc/articles/PMC5825045/ pthalates
    (from plastics) and PCBs implicated in gender dysphoria
https://www.ncbi.nlm.nih.gov/pubmed/28478814 fragrances that
    contain several hormone disruptors linked to transgender incidence
https://www.psychologytoday.com/us/blog/health-matters/201902/the-dissolution-
    gender circumstantial evidence supports hypothesis that endocrine disruptors play a
    role in GD
https://www.sciencedirect.com/science/article/abs/pii/
    S0018506X04002041 Dithylstilbesterol and lesbian orientation, etc.

Sexual Liberation: Whose Sexuality is Liberated?
Does the Sexual revolution cater to men or women?

The evolutionary background:

Gender differences in sexual motivation are accepted among biologists who recognize that females are more discriminating in their choice of mate. From the outset, females invest more in their offspring as the egg is always much larger than the sperm. Among mammals, females also bear the huge energy costs of gestation and lactation. For these reasons, females are considered a resource over which males must compete. Compete they do.

Men's eagerness to mate is highlighted by the sex industries of pornography and prostitution that cater principally to males. In every country studied, men also want more partners in uncommitted relationships although a minority of women are more interested in casual sex than the average man.

Whereas men are generally more interested in casual sex, women look for greater emotional commitment in a relationship. This sensibility is reflected in romance novels, read mainly by women, and it would have helped our female ancestors to select faithful partners who stayed around to help them raise children.

The historical background

As early as 1870, contraceptives (rubber condoms) were widely used in the U.S. and Europe[1]. Before then, extramarital sexuality carried a huge risk of unwanted pregnancy and consequent abandonment. Most women were sexually active only after marriage, keeping the single parenthood ratio to below 5 percent over many centuries of English parish records.

A similar picture applied in the U.S. until the sexual revolution of the 1960s when many young women began having sex before marriage. Sex researchers also document increased interest in sexual pleasure, variety of sexual acts, time devoted to making love, number of sex manuals purchased, and so forth.

What caused the sexual revolution?

Many factors may have been implicated, such as improved contraception (the pill which gave women more control), but effective condoms had been widely used for a century. Marriage prospects and careers were the key. Women's marriage prospects worsened steadily throughout the sixties and there were only 80 men of marriageable age for every 100 women[2] thanks to an echo effect of the baby boom a generation earlier. Women also postponed marriage as they developed careers.

The net result was a large and increasing population of women who were sexually active outside marriage. Facing stiffer competition for men, women upped the ante by offering increased levels of sexual intimacy outside marriage.

In addition to complying with the masculine desire for sex without strings, women today adopt a more masculine sensibility regarding issues of number of sexual partners, sexual variety, and sexual satisfaction.

Which gender is more pleased by those circumstances? Whose evolved psychological needs are being catered to? From an evolutionary perspective, the so-called sexual liberation of women looks more like sexual liberation for men. i.e., men get more sex and more sexual variety without making an emotional commitment.

Because they are over supplied, and less in demand, women enter into the spirit of men's penchant for recreational sex. This psychology is at an extreme on U.S. college campuses where there are only about 75 men per hundred women and hooking up (some level of physical intimacy that lasts for just one night) has largely replaced dating3. As women's bargaining power declines, they must behave more like men if they wish to remain active in the romantic sphere. Women certainly gain in sexual freedom compared to their grandmothers, but they lose out in emotional commitment.

Sources

1. Langford, C. M. (1991). Birth control practices in Great Britain: A review of the evidence from cross-sectional surveys. Population Studies, 45, S49-68.
2. Barber, N. (2002). The science of romance: Secrets of the sexual brain.

Pity the Poor Murderer, His Genes Made Him Do It
Did his genes make him murder?

A criminal defense attorney has many arrows in his/her quiver. The latest is the "warrior gene." Having this gene saved Bradley Waldroup from a first-degree murder conviction.

The charges stemmed from a bloody rampage in which Waldroup shot his wife's friend, Leslie Bradshaw, eight times, killing her, before attempting to kill his wife by chopping her up with a machete.

Waldroup had been drinking as he waited for his estranged wife and their four children who were to spend the weekend at his trailer home in the mountains of Tennessee. When his wife said that she was leaving with her friend, he removed the key from Penny Waldroup's van to ensure that they could not leave, thereby establishing criminal intent. Waldroup then launched his deadly attack on the pair.

The "warrior gene"

Waldroup's defense attorneys ordered a test and established that he had the warrior gene. Like most such biological defenses, there is a germ of scientific truth combined with a hefty dose of junk science, including clever labeling. The warrior gene might be called other things, such as the gambling gene, the depression gene, the irritability gene, or even the live-in-a-trailer gene. Its effects are contingent on an abusive childhood.

The scientific rationale for diminished responsibility is that a variant of the relevant gene, known as MAO-A is linked to the underactive prefrontal cortex, this being a key area of the brain that inhibits antisocial impulses. The gene is also associated with antisocial behavior in European Americans (but not others) but only if they were abused as children[1].

The gene has recently acquired some evidence linking it to impulsive aggression. In an experiment where subjects were provoked by having monetary winnings taken from them, people with the MAO-A variant proved slightly more vengeful, but only if they lost the higher of two amounts of money[2]. They asked for the provoker to drink a larger amount of hot sauce as punishment. Whether this experiment is more relevant to homicidal aggression, or sensitivity to the taste of hot sauce is anybody's guess.

So far, a skilled defense lawyer might weave a tale that the bad gene had gotten the better of the European American defendant. The key scientific problem is that about 34 percent of Europeans have the warrior gene. Yet, homicide is extremely rare at a population level with only about one person in 100 committing a homicide during their lives. If the gene were used to predict homicide, it would be wrong more than 33 times for every occasion that it was right[3].

Just the facts

This brings us back to the Waldroup case tried in March 2009, where the warrior gene formed the kernel of a diminished responsibility defense. This defense received an enthusiastic endorsement in a recent NPR report by Barbara Bradley Haggerty ("Can Your Genes Make You Murder?")

Waldroup's defense was not a simple genetic defense because it was combined with the normally ineffective abuse excuse. Defense expert William Bernet of Vanderbilt University argued that the combination of the warrior gene and being abused as a child was a dangerous cocktail that increased the likelihood of committing a violent offense.

Some of the jurors were persuaded by this defense. According to one, Debbie Beatty: "A diagnosis is a diagnosis, it's there. A bad gene is a bad gene." Junk science is also junk science. There is no getting away from that either, especially if it helps the defense to save a defendant's life.

Sources

1. Crampton, P., & Parkin, C. (2007, March 2). Warrior genes and risk-taking science. Journal of the New Zealand Medical Association, 120 (1250).

2. McDermot, R., et al. (2009). Monoamine oxidase A gene (MAOA) predicts behavioral aggression following provocation. Proceedings of the National Academy of Sciences, 106, 2118-2123.

3. Caspi, A. et al. (2002). Role of genotype in the cycle of violence in maltreated children. Science, 297, 851-854.

Why Are Some Animals Considered Unclean?
Why we avoid eating dogs, cows, or pigs

Religious food taboos are hard to explain. Hindus avoid beef but eat pork. Jews, Moslems, and Seventh Day Adventists eat beef but avoid pork.

It is hard to come up with a satisfactory practical explanation of such conflicting practices, although anthropologists have tried.

Practical explanations

One leading scientific explanation for the Hebraic pork taboo is that pigs make unsuitable domestic animals because they eat everything and therefore compete with their owners for food. This would be an issue in the periodically hot and dry lands of the Middle East where food can be scarce. Contrary to the theory, the taboo persists when Moslems and Jews migrate to more productive ecologies.

The second main theory is that pigs are carriers of trichinosis, a serious parasitic illness. This theory is not too credible either. One issue is that the parasite is destroyed by thorough cooking. A practical answer would be to cook the pork thoroughly rather than avoiding it. Once again, the food taboo persists even when Hebraic peoples migrate to regions where trichinosis is not a problem.

Religious texts are of little help here either. We are told that pigs are unclean because they have cloven hooves, for example, but that is a circular argument. We are not told why one hoof shape is more unsanitary than another.

Such food taboos are intriguing because they are potentially very costly. After all, keeping the taboo can mean the difference between starving and being well fed. According to a relatively new theory of religion, costly food taboos are maintained by their social advantages rather than any practical benefit.

A social explanation

According to the costly signaling theory of religion, people are willing to take on significant ritual costs, such as prayer and fasting, because these practices tie co-religionists together. By paying the ritual costs, a member of certain religions expresses their commitment to the congregation.

Perhaps food taboos are simply another example of a costly ritual. In other words, by identifying with the religion, a member undertakes not only to observe the religious rituals but also to refrain from eating proscribed foods.

The notion that food taboos keep co-religionists together has much going for it. Indeed, it can be rather difficult for people from different religions to share a meal if they are so strict that they fear being contaminated when the unclean food touches their own.

According to this social explanation, it does not really matter what food is proscribed. The important point is that if you are a Hindu you agree to avoid beef without qualification as a sign of your commitment to the belief system. Similarly, Jews agree to avoid pork as one of their taboo species.

Not all food taboos are explicitly religious, of course. In the West, dogs are treated like family members so that any sort of cruelty to them is criminalized. The fact that farmers sometimes resorted to killing and eating the family dog during the Great Depression illustrates how desperate they were for food. In China, India, and other countries, dogs are commonly eaten, however, and restaurants specialize in the preparation of dog meat.

The bottom line, then, is that some animals are tabooed, or eaten, purely because human groups make arbitrary agreements in these matters. If a person wants to fit in, they must follow the taboo. If they do not, they will be seen as different.

Altered States and Other Species
Other species may deliberately alter their interior landscape

We all experience states of mind that can be described in words. Because other species do not have our communicative skills it is easy to assume they lack an interior mental life. That may be a mistake. One key to the mental lives of animals is provided by their vulnerability to addiction.

Pain and Pleasure and Addiction

Neuroscientists discovered that an electrode implanted in the nucleus accumbens will result in experimental animals that become addicted to sensations of pleasure that the electrodes deliver when activated by pressing a lever.

Animals addicted in this way ignore food and other biologically important activities such as mating or attending to offspring. In essence, they behave very much like human junkies.

While other species may be limited in verbal communication, they might still use an artistic channel for communicating an interior life. One sign that animals have internal aesthetic experiences is the existence of their artistic creations, such as paintings.

Painting Primates

Many different species of birds use found objects to enhance their sexual attractiveness to females. For example, Australian bower birds construct a mating bower with grass and twigs and decorate around it with blue objects ranging from berries to pieces of glass. Females choose males having the most elaborate displays, suggesting an appreciation for artistic ability.

Can animals produce creative products similar to those of human artists? Primate researchers provided chimpanzees and other primates with paper, crayons, pens, and other art materials to see what they might generate and how it compared to human artistic products. Early results are presented in zoologist Desmond Morris's fascinating book *The Biology of Art*[1]. The majority of primates generated little more than a jumble of random scribbles. Yet, some chimpanzees are good painters, possibly because of their natural penchant for making, and using, tools that requires fine manipulation of objects. Orangutans and gorillas also produced interesting paintings. A few monkeys made passable works, with highly intelligent cebus monkeys doing the best.

Chimpanzee artists embraced painting with surprising passion. They would stay on task until a particular sheet of paper was more or less filled with color. After they had finished, they would refuse to do any more work, suggesting they had a sense of the painting being completed. A chimpanzee named Congo had a lot more artistic ability than the others. He was the Rembrandt of the chimpanzee world.

The primate paintings were hung in an exhibition without revealing the identity of the artists. Some distinguished art critics were recruited to see what they would make of it. None of the critics suspected the work was by a nonhuman primate, or even by a child. Some believed that the anonymous paintings were the creation of a leading abstract expressionist like Russian painter Wassily Kandinsky.

Anyone who looks at Congo's paintings can understand why they were duped, Superficially, these are blobs of color generally made in the basic fan pattern. Yet they have their own distinct esthetic with a sensitive choice of colors and an unmistakable feeling of balance, control and restraint. They are hauntingly beautiful, suggesting an interior world of aesthetic sensitivity.

In addition to such glimmerings of a capacity for aesthetic experiences, chimpanzees raised with humans evince a remarkable similarity of interests and activities.

Bonobo Picnics

One remarkable demonstration of this was provided by Kanzi, a bonobo, as described by George McGavin in the 2014 BBC series Monkey Planet. Kanzi ordered food for a picnic over a cell phone. He did this by pressing panels on a computer screen that called out the names of the foods. At the picnic site, he set a fire, using a large match to start the fire. Then he speared marshmallows on a stick and turned them over a flame until toasted the right amount.

It is hard to watch this clip without getting the impression that the deliberation and planning involved in Kanzi's picnic could not happen without interior experiences very similar to our own.

Another sign of having subjective experiences is provided by voluntary activities aimed at altering internal states.

Psychedelics

Attraction to mind-altering drugs is found for raccoons that ingest alcohol in fruit that ferments in the summer heat. Chimpanzees were seen to enjoy smoking marijuana in the company of their trainers in a more laid-back period of academic history[2].

The behavior of black lemurs in relation to toxic millipedes is also instructive, as described by George McGavin in Monkey Planet. The lemurs bite the bugs, inducing them to release poisons. Then, they rub the millipedes all over their fur.

This odd behavior is not well explained but the animals enter a state similar to human intoxication. Apparently, they are motivated by a human-like need for variety of experience or - dare one say - altered states of consciousness.

Of course, many hard-nosed scientists would argue that we can have no idea about the mental states of other species. This once-confident assertion is vulnerable to research in neuroscience suggesting that pleasure, anger, and disappointment, are registered similarly in the brains of monkeys and humans.

Sources

1. Morris, D. (1962) *The Biology of Art.* New York: Knopf.

2. Project Nim (2011, July 20), *"Project Nim: A Chimp's Very Human Very Sad Life,"* http://www.npr.org/2011/07/20/138467156/project-nim-a-chimps-very-human-very-sad-life. https://www.atlasobscura.com/articles/lemurs-get-high-on-their-millipede-supply

Are You a Sucker?

Who gets conned and why

Social psychologist Robert Cialdini is an expert on manipulative communications[1]. Yet, he was the first to admit that he was more vulnerable than most. He was the man accepting the "free" rose from a cult member when accosted in the street.

Cialdini recognized that the key to most influence strategies is to make people feel good about doing what you want them to do. He describes a scary human capacity to agree, to obey, to accept absurd religious claims, and to avoid reasoning and critical thought.

Anyone who stops to give the man on the corner with the "Homeless: Will Work For Food" sign five dollars probably feels good about helping the destitute. They would feel less good if they realized that the man with the sign made more from pan handling than they did from work.

The modern world is full of opportunities to get ripped off by confidence tricksters and most of them are not working the streets. Have you succumbed? Let me list some of the ways it could have happened whether you were aware of it or not.

You may be a sucker if ...

There is a great variety of feel-good tactics deployed by influence practitioners. One of the most prevalent is the free gift/sample. It is an axiom of marketing that there are no truly free gifts. One reason is that most people have a strongly developed sense of fairness. That means that when you accept a gift, you incur a psychological obligation to repay. That often means purchasing something you do not need.

Otherwise, free gifts may be repaid by giving up personal information, right down to credit card numbers. Take the one-month free trial offered by Netflix. This is free only in a monetary sense. It is up personal information that is far more valuable to Netflix than the one-month of free access. After all, the access costs them virtually nothing. Yet, the information they receive can be parlayed into a long-term subscription worth hundreds of dollars.

Free gifts are not always concrete like the food samples given out in supermarkets, or of monetary value like a subscription. They may be social. There is a reason that salespeople are often selected for their physical attractiveness. Attractive people boost the confidence and happiness of people with whom they interact. Their free gift is sex appeal. Sex sells because it is pleasant. When people feel optimistic, they are more likely to make purchases – even large and impulsive ones.

The manipulation of beauty in advertising is so obvious that it is hardly worth mentioning. Yet, there are intriguing wrinkles. One is that the models used to sell luxury cars are often sexy young females even though many of the customers are middle-aged men. The implication is that in this car the man will appeal to sexy women[2].

Such ads were first created by disgraced behaviorist psychologist John B. Watson who was inspired by Pavlov's work with salivating dogs. The idea is that if the luxury car is paired with the sexy woman repeatedly, the car will elicit the same excitement as the woman.

So much for sexually salivating men! What do such ads have for women who are buying more and more luxury cars? Women likely identify with beautiful females and get something of a lift, as though they were that attractive themselves. Otherwise, it would be hard to explain why ads in ladies' magazines contain so many beautiful women. If men are being convinced that the product will get them beautiful lovers, women are being convinced that the product will make them highly desirable. It may be a false promise, but most people are swayed by sexual advertising without even being aware of it[2].

Advertising exploits the sex appeal of women's bodies which are featured more often than men's bodies[3]. Advertisers are also attuned to refined messages transmitted by female body shape. Models selling to women are mostly slender, evidently catering to the feminine desire to lose weight[2]. Models selling to men are generally more curvaceous. Beer commercials that are targeted to comparatively low-income men feature large-breasted women[3] that are believed to appeal to this demographic, for example.

So, we are all suckers in one way or another, allowing ourselves to accept as fact claims that make us feel good even as we recognize that they are deceitful, or wrong. Just because we resist the obvious frauds like e-mails promising to wire a hoard of money into our accounts, we get taken anyway.

Sources

1. *Cialdini, R. B. (1988). Influence: Science and practice. Glenview, IL: Scott Foresman/Little.*

2. *Reichert, T. (2003). Sex in advertising: Perspectives on the erotic appeal. Mahwah, NJ: Lawrence Erlbaum.*

3. *Lijima Hall, C. C., & Crum, M. J. (1994). Women and "body-isms" in television beer commercials. Sex Roles, 31, 329-337.*

U.S. Body Politic Mounts Immune Defense
The pandemic coincided with an assault on democracy. The defenses are similar

The assault of a pandemic has been accompanied by an assault on democracy: the defenses are similar.

At the peak of a pandemic, we are acutely aware of the body's immune response to disease. Something similar is occurring in the political response to the January 6 insurrection at the Capitol.

Some of the responses to an attempted coup are similar to the immune system's response to a pathogen.

The immune system has an inherited, or primary, response to disease that involves physical barriers, like skin, in addition to specialized defensive cells such as macrophages and neutrophils.

The immune system also has an adaptive response in which the immune system acquires a capacity to recognize pathogens, and to inactivate them.

How a State Defends Itself Against Inside Attacks
Pathogens are not necessarily from outside the body. They sometimes attack from within. This is true of autoimmune diseases, for example, where the body mounts an immune response against its own tissues.

The U.S. insurrection was clearly of this type. The conflict has been brewing for years in efforts to deceive the public through misinformation on social media. Before the 2020 election, the sitting president said that he would not accept electoral defeat.

This was the culmination of a campaign of deception in which the president insisted that the election had been "stolen." Inciting a mob to attack the Capitol as lawmakers formalized Biden's electoral victory was more of the same.

Most contemporary insurrections come from within the country. The inherited response of a state to insurrection comes in the form of laws and institutions that set the rules of a democracy.

This defense was tested repeatedly in numerous unfounded challenges to the legitimacy of election results. The Secretaries of State of all 50 states came through with flying colors as did each of the judicial branches, from the Supreme Court down, in which these baseless claims of fraud were adjudicated.

The shakiest line of defense turned out to be the Congress itself, where large numbers of conservative lawmakers voted to reject electors under threat of losing their seats through the actions of a vengeful party leader.

When all else fails, the state uses its military power to protect its institutions, as reflected in the 25,000 troops defending the Capitol during the inauguration in addition to military protection of all state capitols.

While U.S. democracy survived the attempt to pursue an authoritarian state, the enemy remains within, like an invading pathogen and elements of the government are mounting an adaptive (i.e., learning-based) response in which they seek to identify and neutralize the enemy within the body politic.

The Adaptive Response

Every new threat to a state is different from previous ones so that all arms of government, including the judicial system, the legislature, and the executive must learn new ways of countering it.

This response involves collecting information about the enemy. The immune system is similar and macrophages recognize new pathogens using pattern recognition proteins. This recognition process facilitates the entry of defensive neutrophils to infected tissues.

The Trump-led assault on democracy began, as other totalitarian efforts have, with a propaganda campaign that sought to divide the nation and create political and racial divisions, turning political opponents into the enemy. This campaign ultimately brought the absurd extreme of conspiracy theories of the QAnon type.

The radicalization campaign led directly to the Capitol insurrection and was mainly fought via social media where all manner of antisocials could wallow in a community based on hate-filled falsehoods spurred on by the president.

The temperature of these hate-based rants was reduced by social media companies who de-platformed Trump and others. This result is the first sign of a functioning defense of the body politic even if it came from companies rather than Congress.

Democracy may have won this battle, but we can expect a long war in which a large minority of the population has been radicalized through a diet of internet lies. The enemy is within the body politic but the adaptive response has also begun.

Why Conservatives Spend More on Cyber Porn
It does not stop them trumpeting their morality

Use of pornography by political and religious conservatives is intriguing because most are so strongly opposed to it. Recent research concludes that Christian college students are somewhat less likely to use online pornography than the general college population but still found that the "majority of males had some involvement in Internet pornography".

It is hard for religious people to admit to behavior of which they so strongly disapprove. This introduces unreliability into such self-reported data. A Harvard study provided objective evidence about actual consumption of online pornography by paying customers. This study found a consistent pattern of more conservative, and more religious, states spending more on pornography.

The biggest consumer of Internet pornography was Utah with 5.47 subscriptions per thousand home broadband users compared to Montana, the lowest state, with 1.92 subscribers per thousand[1]. Study author, Benjamin Edelman of Harvard Business School, focused on broadband users because pornography is a bandwidth hog. Edelman was also careful to rule out the age distribution of the population, income, education, population density, marriage rates and other characteristics that might make state comparisons unfair. Utah still wound up at the top of the heap.

Utah's top-ranking surprises many. One can think of different adjectives to describe the state: religious, conservative, family-oriented, outdoorsy, clean-living, but few would have guessed top pornography-consuming. Many would find it easier to attribute such interests to western neighbor Nevada, a center for bricks-and-mortar gambling and prostitution. Nevada didn't even make it to the top ten.

States that banned gay marriage (implying conservative political views) had 11 percent more porn subscribers. The level of agreement in a state with the statement that "Even today miracles are performed by the power of God" predicted higher pornography consumption. States claiming to have old-fashioned values about family and marriage purchased substantially more adult-content subscriptions. Eight of the top ten pornography consumers were red states in the last Presidential election (Utah, Alaska, Mississippi, Oklahoma, Arkansas, North Dakota, Louisiana, and West Virginia) with blue states, Hawaii (#4) and Florida (#9) bucking the trend.

In addition to the conservative states' avid consumption of Internet pornography, there have been numerous examples of prominent conservative politicians and public figures whose lofty statement of values in sexual matters was cruelly undercut by their own actions: Anthony Wiener, Larry Craig; Newt Gingrich; Mark Foley, Bob Livingston, Henry Hyde, and Bob Packwood, among scores of less recognizable names. It is not just conservative politicians who fail their own moral tests. The parade of prominent evangelical preachers similarly disgraced, including Jim Baker, Jimmy Swaggart, and Ted Haggard, is equally striking. Episcopal minister Marshall Brown was recently fired for accessing online pornography at work.

Many words have been used to explain the apparent contradiction between ideals and practices. Hypocrisy is the obvious one. Accessibility is another issue and the illusion of anonymity that the Internet offers. Edelman cites repression, pointing out that if people are told they cannot have something they want it more.

Although his findings might appear new and shocking, not much is genuinely new under the sun. Many decades ago, sociologist Laud Humphreys[2,3], author of the book *Tearoom Trade*, wondered what kind of men would stop off in a public restroom for a few minutes of casual sex with other men, on their way home from work. He jotted down their car license numbers and tricked the local motor vehicles department into divulging the men's addresses. Without mentioning the true intent of his study, Humphreys interviewed the men in their homes. Most seemed happily married. Their homes often had the U.S. flag on the wall and a Bible on the mantelpiece.

Humphreys had the impression that their aura of respectability was overdone. He referred to this as the "breastplate of righteousness," or a defense against accusations of sexual impropriety by seeming very righteous.

The bottom line, then, is that however much conservatives trumpet their sexual morality, they are not better than the rest of us. Indeed, the evidence suggests they are slightly worse.

Sources

1. Edelman, Benjamin (2009). Red light states: Who buys online adult entertainment? Journal of Economic Perspectives, 23, 209-220.
2. Humphreys, Laud (1970). Tearoom trade: Impersonal sex in public places. Chicago, Aldine.
3. Barber, N, (2002). Encyclopedia of ethics in science and technology. New York: Facts on File.

Ladies in Heat?

Scholars argue that women have estrus or sexual heat

The Romans considered that female mammals were driven mad by sexual desire. The term "estrus" describing female sexual heat is derived from the word "oestren" or gadfly. Cows race madly around their pastures to escape being bitten by the gadfly. A recent scholarly book argues that women also experience estrus.

This is something of a revelation given that the received wisdom of sex researchers was that women had lost estrus entirely or had "concealed" it to permit continuous sexual receptivity. Being sexually receptive all of the time kept men in the dark as to when ovulation was actually taking place. This meant that if a man were to impregnate a woman, his chances were improved by staying around rather than just showing up during the few days of the month when a woman is most likely to conceive.

In making the case that women have estrus sexuality[1], Randy Thornhill and Steven Gangestad of the University of New Mexico had to hack away at what they see as a false stereotype of sexual heat in females. Contrary to the view that females in heat are indiscriminate, Thornhill and his longtime collaborator argue that the opposite is true. They conclude that although female sexual interest is heightened at estrus, they become highly discriminating in their choice of a mate.

Female chimpanzees mate with all the adult males in their group when their sexual swellings appear signaling impending estrus. When they are likely to conceive, however, they disappear with a single high-quality consort male who will father their offspring.

Changing views of female sexuality

Prior to Darwin, females were considered to have inconsistent sexual interests, if they had any at all. Darwin proposed that female choice is an important mechanism in evolution, resulting in such phenomena as the peacock's tail, but others were skeptical. A century passed before Darwin's ideas on sexual selection were taken very seriously. Researchers began to amass evidence of male adaptations designed to impress choosy females, such as the gaudy plumage of male birds.

In addition to hooking up with men having good genes, women also turned out to be picky about the social prospects of potential mates. They preferred men who were socially successful, had valuable skills, enjoyed high social rank, or wealth. A willingness to invest in children was also valued.

Thornhill and Gangestad believe that women have essentially two different kinds of sexual responsiveness. That occurring around ovulation is referred to as estrus sexuality. Sexual responsiveness outside the fertile phase of the monthly cycle is referred to as extended sexuality. They believe that estrus sexuality is designed to obtain good genes for offspring whereas extended sexuality involves those other benefits, e.g., food, that women may extract from men by being sexually receptive when they are not fertile.

Estrus sexuality

Given that women are sexually receptive for most of their monthly cycle, estrus cannot be defined by receptivity alone. This means that the greatest burden for Thornhill and Gangestad in establishing estrus sexuality for women is placed on the argument that women are more discriminating in their choice of a mate during this phase of their cycle.

They found a remarkable amount of evidence in support of the view that women are far choosier around the time of ovulation. In particular, during estrus women are more attracted to men with highly masculine faces and bodies. They are also more swayed by bodily symmetry, which is a reliable index of genetic quality. They are even more attracted by the scent of symmetrical men. In estrus, women find highly masculine voices more attractive as well.

Like other mammals, it seems that women are choosier about what sort of man they will mate with when they are most fertile. This implies that their sexual psychology is designed to obtain good genes, rather than simply to secure any sperm.

Just when we feel that we are finally sorting out the difficult topic of human female sexuality, we discover that it is still more complex. If Thornhill and Gangestad are correct, then when women become more sexually liberated in some societies, this would be due to the benefits of sex outside estrus. Yet, women in sexually liberated societies are both more interested in male physical attractiveness (suggesting estrus[2]) and more interested in having various partners (suggesting extended sexuality). There is not a watertight distinction between getting good genes and mating for non-genetic reasons after all.

Sources

1. Thornhill, R., & Gangestad, S. W. (2008). *The evolutionary biology of human female sexuality*. New York: Oxford University Press.
2. Swami, V., & Tovee, M. J. (2005). *Male physical attractiveness in Britain and Malaysia: A cross-cultural study*. Body Image, 2, 383-393.

Choreography for the Birds
Order in nature is often less intelligent than it appears

When Darwin was a young man, religious people took to the hills, fields, streams, and moors to pursue enthusiasms for birds, butterflies, and flowers[1]. Appreciation of nature was a way to praise the creator.

We should not be too surprised that Darwin was affected by this religious enthusiasm. After all, he fully expected to take up his occupation as a clergyman. Natural theology highlighted the exquisite match between each species and the way that it makes its living, something we now refer to as adaptation.

Watchmaker and Watch

One of the most influential natural theologists of the time was William Paley who promoted creationism using clever rhetorical turns that are superficially convincing but lack logical validity[1].

One of Paley's favorite metaphors was of the creator as a watchmaker, and the natural world as a clock. If one comes across a beautiful work of engineering, such as a fine watch with its delicate springs and escapements, it is obvious that there has to be a watchmaker, he argued. (The case that design in nature calls for a creator is referred to as the argument from design).

While this argument is pure sophistry, it has not prevented contemporary creationists from repeating it in various forms. To begin with, the argument is circular because we already knew that watches are made by watchmakers and the inference that this would apply to something completely different is bogus. Any conclusion about man made objects has zero relevance for living organisms that are not man-made.

Such deductive errors can be surprisingly compelling, a fact that casts doubt on the robustness of human reasoning. Although logically flawed, such arguments from design are emotionally satisfying and the emotion often wins out[2].

Whereas natural theology used the natural world as supportive evidence for theological claims, Darwin's theory of evolution undercut such claims. If natural selection could explain why animals are ideally matched to their ecological circumstances, then it removed the benevolent creator from the picture. In other words, the "why" of theology was overtaken by the "how" of natural selection. With the mechanism of natural selection in place, there was no need for a creator.

Natural theology may be a footnote in history, but it illustrates a strong human tendency to ascribe central agency and control to phenomena that may lack either. Coordinated movements of social animals are a case in point.

The Dance of the Starlings

Huge flocks of starlings can wheel in unison around the sky in coordinated movements that recall the terrestrial feats of human choreographers. Dance troupes, however, require detailed drilling and practice, and use a musical score to keep in time. How do starlings achieve such order in large flocks?

Starlings are not alone in their coordinated group movements. Schooling fish do something similar beneath the surface of the water[2]. Acting as a unit in this way may confuse or intimidate potential predators. Or it could simply be a manifestation of the principle that there is safety in numbers. Members of a large school are less likely to be taken in an attack. Isolated fish are more vulnerable.

Either way, the underwater ballet is an impressive phenomenon from the point of view of individuals timing their movements perfectly to match the direction and speed of the group.

Such accomplishments were a mystery, given that there is no one calling the dance, so to speak. No centralized mechanism of control applies. Now, it is becoming clear that the coordination is an emergent property of the actions of individuals, i.e., the coordinated movement of large numbers of individuals rather than the result of centralized control[2]. They are said to be "self-organized."

Fish have sensitive pressure detectors on the sides of their bodies, known as lateral line organs, that are exquisitely sensitive to local disturbances in the water and allow individuals to adjust their moves to avoid colliding with neighbors. Similarly, birds use mostly visual cues to avoid midair collisions. In each case, the result is an illusion of centralized control where none actually exists.

The Human Analog of Starling Choreography

Although we now know that there is no starling choreographer, no lead bird, their aerial acrobatics create a strong impression of top-down leadership and control.

Human cognition is generally flawed by the tendency to attribute agency when none exists. Indigenous people worship thunder as a deity and many see thunderstorms as a symptom of celestial anger.

That cognitive bias likely exists because of the need to distinguish friends from foes. We are therefore very willing to ascribe motives to others that may not be accurate, but doing so can help us to form alliances and avoid trouble.

Just as Darwin's contemporaries were happy to ascribe the match between species and their way of life to the actions of a benevolent creator, we are still inclined to see the natural world as a great deal more intelligent, and more deliberate, than it actually is.

A striking example of this emerged in modern evolutionary biology when scholars began talking about "selfish genes[4]," an unfortunate metaphor that cast a cloud of confusion over a generation of students and scholars. In reality, genes do nothing more than supply the recipe for protein molecules and are expressed under restricted biochemical circumstances[5]. They cannot be either selfless, or selfish.

Even if starlings need no director to wheel in unison through the sky in vast flocks, it remains an impressive spectacle. We can all agree to enjoy the show.

Sources

1. Ruse, M. (1982). *Darwinism defended: A guide to the evolution controversies.* London: Addison-Wesley.

2. Kahneman D. (2011). *Thinking fast and slow.* New York: Farrar, Straus and Giroux.

3. Hemelrijk, C. K., and Hildenbrandt, H. (2012). *Schools of fish and flocks of birds: Their shape and internal structure by self-organization. Interface Focus, 2,* 726-737.

Why Cults Are Mindless
Mindless obedience keeps religious cults together

Whenever a cult clashes with the law, the public is fascinated and horrified by the capacity of leaders to control members. Perhaps the members surrender all their property. Or they are sworn to celibacy, leaving only the leader with sexual access to all the women. This was the case with David Koresh, leader of the Branch Davidians, the cult destroyed in a Waco, Texas, fire[1]. Or cult followers drink poison on command as in the Jonesboro, Guyana, tragedy.

The secret recipe of all such cults may lie with the members rather than the leaders. Social psychologists discovered that members get overly attached to cults that ask a great deal of them.

When a lot is asked ...

Research on U.S. communes suggests that organizations need to be quite demanding to get their members committed enough to stay the distance. When sociologists Richard Sosis and Eric Bressler[2] studied 83 19th-Century communes in the U.S., they found two intriguing patterns. The first was that more demanding communes lasted longer. Bigger sacrifices engendered greater emotional commitment to the group.

The communes could be extreme. Some required vegetarianism, or celibacy, or surrender of all worldly possessions to the collective. The more demanding a religious commune was, the greater the level of cooperation it elicited from members and the longer the community survived. Groups with fewer than two costly requirements lasted less than ten years, on average. Communes that had between 6 and 8 burdensome costs lasted for over 50 years and those with more than 11 were still in business after 60 years.

Costly commitment helped groups stick together only in religious communities. This offers a fascinating glimpse into the socially cohesive function of supernatural beliefs.

Secular communities were particularly unstable, generally lasting less than ten years. Contrary to the pattern for religious communities, the more demanding secular communities folded more quickly. Indeed, the most demanding secular community closed its doors after only a year.

Why was there such a difference between religious and non-religious communes? Evidently, sacrifices made for the community are interpreted differently for members of religious communes compared to secular ones.

Ratcheting up the costs of membership works well only for religious communities. A supernatural belief system can justify heavy membership costs in terms of a higher purpose. Without supernatural justification members might ask themselves why they are paying so much to be in the commune. Lacking a religious justification, they may conclude that they are being exploited by the leadership. The logical solution is to leave.

When they are backed up by a religious belief system, communes can tolerate considerable inequality. This may be illustrated by differences in permissible sexual behavior.

Sexual inequality

A secular commune requiring celibacy from all male initiates would be destabilized by the free sexual expression of the leader.

Yet, that sort of inequality may work if members believe that the leader is a divine incarnation. Something close to this scenario played out in the David Koresh cult (the Branch Davidians), that was wiped out in a fire following a standoff with federal authorities near Waco, Texas, in 1993[1]. Evidently, Koresh had free sexual access to female members consistent with his divine status whereas other men were expected to be celibate[3].

So, religious cults that survive for more than a few years are characterized by blind obedience. The really difficult question for relatives, clinical psychologists, and researchers is why inductees are so willing to surrender their autonomy in the first place.

Yet, cults are not unusual in this respect. Mindless obedience to authority figures is evident in many other settings. These include the army, Greek societies, sports organizations with their Byzantine rules, business corporations with their company men and women, and the groupthink of political life. World religions can also be included, of course.

Mindless obedience is good for the cult, but it is generally not good for the member. The same principle applies to entire countries. The less inequality there is, the better the quality of life experienced by everyone[4].

Sources

1. Barber, N. (2012). Why atheism will replace religion: The triumph of earthly pleasures over pie in the sky. E-book, available at:
http://www.amazon.com/Atheism-Will-Replace-Religion-ebook/
dp/B00886ZSJ6/
2. Sosis, R., & Bressler, E. (2003). Cooperation and commune longevity: A test of the costly signaling theory of religion. Cross-Cultural Research, 37, 211-239.
3. Newport, K. G. C. (2006). The branch Davidians of Waco: The history and beliefs of an apocalyptic sect. New York: Oxford University Press.
4. Wilkinson, R., & Pickett, K. (2010). The spirit level: Why greater equality makes societies stronger. New York: Bloomsbury Press.

The Commons Is Not a Tragedy, But Markets Can Be

Many shared properties (commons) flourished for centuries and still do

G arret Hardin's classic analysis of commons grazing systems concluded that they are inherently dysfunctional and that they disintegrate due to human greed[1]. Free-market theorists love this story as a justification for market systems. It turns out to be completely fictitious.

In commons systems, communities divide valuable resources that they hold in common. In English feudal villages, each villager typically grazed a milking cow in the shared pasture. These traditional grazing arrangements continued successfully for many centuries[2].

The same conclusion applies to many other historical commons systems whether devoted to sustainable grazing in Mongolia, communal water management in arid regions, lobster conservation in Maine, forest management in Nepal, or other purposes. Of course, the digital commons as illustrated by Wikipedia, or open-source software, is a model of how freely shared information profits everyone.

Despite the historical fact that commons systems are successful, Garret Hardin set out to explain why they could not work in theory. His analysis was so seductive to economists and social scientists that they conveniently ignored the historical evidence.

This is the equivalent of a contemporary thesis claiming that people cannot fly in a craft made of metal.

The "Tragedy" Scenario

Hardin's analysis of commons systems highlights the conflict between what is good for the common grazing land and what is good for the individual herder[1].

In a commons grazing system, putting out an extra animal on the shared land is always in the interests of the individual herder. However, if

144

everyone does this the land gets overgrazed. Bare patches are produced and weeds begin to take over. The overgrazed land deteriorates and loses its value.

Hardin argues that this outcome is inevitable because the benefits of an additional animal to the herder always outweigh his marginal cost in terms of lost grazing quality. This conclusion is depressing and inevitable, hence a "tragedy."

There is no doubt that this tragedy may unfold in many different scenarios from throwing litter on a roadway, to hogging a good parking place. In these examples, public resources, or commons, whether pristine roads or convenient parking, are degraded by the selfish actions of individuals. One way of solving the problem is to introduce a fee for parking, i.e., a market-based solution.

Commons grazing systems persisted for centuries in rural societies, which is Hardin's key example. But commons systems also thrive in modern societies as cooperative businesses from utility companies to food co-ops[2,3].

Medieval English villagers operated stable commons farming. No one got greedy and no one decided to graze an extra cow. The reason was that when an extra cow was brought home for milking, her presence would be obvious to all the other villagers. In that environment, social pressures were powerful and might be backed up with violence. Interpersonal disputes were solved with knives used in eating. People who were stabbed rarely died of their wounds so much as the ensuing infections[4].

The commons was stable because no cheating was tolerated. Ironically, the very thing that destroyed the commons was not the selfishness of the herders but the rapacity of markets, specifically the market for land.

145

Why English Commons Grazing Ended: Enclosure, Dispossession, and Dislocation

The end of widespread commons grazing in England was not due to any inherent failure. It was not that sort of tragedy. Instead, it is an example of laissez-faire capitalism at its most predatory.

The land enclosure movement was one of the biggest appropriations of common property by rapacious market forces. Millions of peasants were forcibly removed from their homes, deprived of a livelihood and forced to seek refuge in dodgy urban slums.

Land was enclosed because it had become more valuable. Landowners could make far more money by selling agricultural produce in towns and cities than they could glean from the limited produce of their resident labor force who resembled impoverished sharecroppers in the rural South of the U.S.

The grim fate of English peasants furnishes one example of the socially destabilizing impact of markets. This tragedy of the markets is wreaking its havoc today.

Markets Are Not Stable

Contrary to the widely articulated view that free markets make everyone wealthier and happier, they generally foster unforeseen dystopian outcomes that are "tragic" in just the ways, and for the same reasons, that Hardin falsely attributed to commons systems.

Even Adam Smith, the father of free-market economics, foresaw many problems with unregulated markets[5]. He wrote that whenever you get a handful of merchants together in a room, the first topic of conversation is how they can fix prices.

One of the most obvious problems in our contemporary economy is wages being fixed at very low levels for unskilled workers. The problem reflects a poorly regulated, newly internationalized, labor market.

Companies can reduce their labor costs by moving plants to low-cost countries. Or, they may outsource labor directly by having employees log in remotely from distant places, like India.

All Exchange Systems Require Rules and Enforcement

Wherever one looks, unchecked market systems create instability and exact a price in human misery. Conversely, commons systems can generate great stability on a time scale of centuries.

When commons systems fail, it is generally because the commons property is appropriated by market systems. Surviving commons grazing systems occur mainly in marginal land, such as hill pasture, or arid scrubland. This is because valuable land succumbs to the hungry market. It is not due to the greed of the commoners.

Whether we derive sustenance from the marketplace, or from a commons system, we participate in a social exchange process either by drawing on shared resources in an agreed-upon manner or purchasing goods in a market.

Successful, long-lasting commons systems, like those in Switzerland, have two main characteristics. First, they are immune to takeover by markets[3]. Second, they operate according to consensual rules whereby cheats are punished by fines. The commons is not a tragedy. The real tragedy is unregulated markets.

Sources

1. Hardin, G. C. (1968). The tragedy of the commons. *Science, 162*, 1243-1248.

2. Curl, J. (2009). *For all the people: Uncovering the hidden history of cooperation, cooperative movements, and communalism in US history*. Oakland, CA: PM Press.

3 Ostrom, E. (1990). *Governing the commons*. Cambridge, England: Cambridge University Press.

4. Tuchman, B. (1978). *A distant mirror: The calamitous 14th century*. New York: Alfred A. Knopf.

5. Smith, A. (1776/2015). *The wealth of nations: A translation into modern English*. Manchester, England: Industrial Systems Research

Why Weather Affects Mood

Weather extremes are demotivating in the short term...

Even as many of us were trapped indoors due to the pandemic, our moods rise on bright days and plumb the depths after days of rain. Such transitory mood swings are not well understood, but animal behavior offers useful insights.

Adaptive Inactivity

Low moods are demotivating. We feel lethargic, or tired, and are less likely to exert ourselves in new or productive activities. While we may feel listless, dispirited, and unhappy, this could boost the biological currencies of survival and reproduction.

We wake up when it gets bright in the morning so our activities follow a rough circadian rhythm, as do internal physiological systems from digestion to immune function. For example, we eat less often during the night because insulin production slows and less sugar is withdrawn from the bloodstream, which blunts hunger.

Knowledge about the restorative value of sleep is expanding by leaps and bounds, from its effects on immune function to memory to flushing the brain of impurities[1]. Feeling tired encourages us to lie down and sleep each night.

The same sort of circadian rhythm is manifested in most other vertebrates although some, like owls, become active at night. Cycles of rest and activity are surprisingly flexible and shift workers are forced to reverse the usual pattern which can impose health costs. Similarly, species that are normally diurnal can shift to becoming nocturnal due to human influence, as is true of coyotes, boars, elephants, and tigers, who avoid unwanted human attention by becoming active at night.

There are strong seasonal patterns for species that rest more in winter. Some go to the extreme of hibernating when they sleep in a warm den through much of the cold season. In spring, longer days rouse hibernators from their slumber and motivate some species to embark on their seasonal migration.

Light and Reproduction

These effects are illustrated in the way day length alters the hormonal status of seasonal breeders like birds. With increased day length, migratory species become increasingly restless and head off in the direction of their seasonal migration.

In spring, as day length increases, male birds experience a surge in testosterone, fight over territories, and defend them with the territorial song. Singing is itself related to testosterone as the song controlling structures of the brain wax and wane with the breeding system.

Humans are not seasonal breeders, but our brains also have complex responses to day length. These are highlighted by the phenomenon of wintertime depression or seasonal affective disorder, that occurs in highly seasonal places distant from the equator where day length in winter is very short, triggering severe depression in vulnerable individuals.

Although we are not a seasonal species, our reproductive systems are affected by the length of day. Interestingly, men have an annual cycle in circulating testosterone levels that peak in autumn.

The Seasonal Shift

Most of us tolerate the short days of winter, but being confined at home due to extreme cold interferes with our customary activities and thereby lowers mood.

When temperatures warm up in summer, we spend more time outdoors and are more physically active whether this involves sports activities, exercise, or outdoor hobbies like gardening. Most people prefer milder temperatures and express greater feelings of optimism and joy.

One side effect of being outdoors longer is that violent crime rates go up. The fact that most civil unrest occurs in midsummer used to be attributed to extreme heat increasing irritability and aggression, but there is no evidence of this. If anything, extreme heat makes us unwilling to move around, much less attack anyone.

The demotivating effect of extreme summer heat is similar to that of extreme winter cold. Both are stressful, tending to increase anxiety and lower mood.

Stress and Precautionary Behavior

Extreme winter cold and extreme summer heat crimp everyday activities in similar ways. In cold climates, there is a temptation to stay indoors more and to get less exercise. When a person does brave the elements, they must spend time putting on extra winter gear that has to be removed on their return. This is a time-consuming annoyance that we do not have when the weather is mild.

When the weather is hot, we have the trouble of putting on protective cream to prevent sunburn and wearing sun hats. Humid conditions are highly uncomfortable and demotivating. This means that we spend more time in air-conditioned buildings and vehicles and spend less time outdoors.

Whether hot or cold, extreme conditions can be highly unpleasant and demotivating. We must also spend a lot of money to control the environment in our homes. All these aspects of extreme weather make them potentially stressful, anxiety-provoking, and depressing.

Despite this, there is surprisingly little evidence that climate has any reliable impact on mood or mental health. There is a very simple explanation for this which is that we are good at adapting to the particular conditions in which we find ourselves.

Residents of Minnesota are accustomed to long, hard winters and many enjoy spending time outside pursuing winter activities such as skating and ice fishing. They are accustomed to the extreme cold and take it in their stride. There is also some physiological adaptation with greater bodily heat production after prolonged cold exposure.

So, we encounter a strange paradox in which harsh weather is ostensibly stressful but has little obvious impact on mood because we are good at adapting to varied environmental conditions. This trait may be what allowed our ancestors to occupy Europe and Eurasia when they were in the grip of an ice age.

Source

1. *Walker, M. (2017). Why we sleep: The new science of sleep and dreams. New York: Penguin/Random House.*

The Many Reasons Why People Have Sex

Humans are more likely to be sexually active when they cannot conceive

Sexuality is about a lot more than reproduction. For our species, by far the most sex happens in contexts where conception is impossible.

Humans are not alone in having sex for many other reasons. When one looks at closely related species, non-reproductive sex is also common.

Other Apes

Female chimpanzees mate with all of the males in a group, for example, although they behave more selectively around the time that fertilization occurs.

Pygmy chimps, or bonobos, are the great-ape champions of promiscuous sex and certainly put humans in the shade in this respect. Every adult is liable to rub genitals with just about any other. Why is there so much sexual activity that is clearly non-reproductive? How does this relate to humans? For bonobos, the purpose of most sexual interaction is clearly social rather than reproductive.

Evidently, sex serves to diffuse aggression and reinforce social bonds. Humans are primarily monogamous, so these functions apply primarily to married couples.

An active sex life helps to reinforce sexual bonds, and couples may make love following an argument as one way of reducing conflict. This might help to explain why there is so much human sexuality outside of the context of reproduction.

Non-Reproductive Functions of Sex

There is an almost bewildering variety of reasons for non-reproductive sex in humans and other species. The extraction of resources (generally from males by females) is observed in several species. Many species of birds and insects provide a nuptial gift of food as a

The human analogy involves prostitution, which predates money and is referred to as the oldest profession.

In forager (i.e., hunter-gatherer) societies, good hunters enjoy more extramarital relationships, and their trysts may be preceded by a secret gift of meat.

Among chimpanzees, promiscuous mating by females before their time of ovulation probably has more to it than meets the eye. In particular, it is thought to be a defense against infanticide. The reasoning goes that males, who typically kill young whenever they can, are less likely to do away with the offspring of females with whom they have mated, thereby destroying their own progeny.

Infanticide is common in some forager societies, such as the Ache of Paraguay[1]. A widow may lose her youngest children when she remarries, because the new husband expects her to invest in his offspring right away.

Among humans, consensual extramarital sex is surprisingly common. Such informal polyandry, or "wife-swapping," is particularly common in societies where mortality rates among men are high[2]. Such favors were extended to guests in traditional Eskimo societies and incurred an obligation to repay in kind. Moreover, in the event that a man died, his guest assumed responsibility for his children.

However arcane the functional explanation may be, most people assume that the prevalence of sexual behavior outside reproductive contexts is simply motivated by pleasure. This is always part of the story, but rarely excludes other possibilities.

Homosexual behavior among humans strengthens alliances, as illustrated by the customary relationships of warriors in ancient Greece, where an older man trained an apprentice and slept with him. Homosexual interactions are surprisingly common for many other species, although it is unclear whether this serves any evolutionary function[4].

Sexual Pleasure: Immediate Versus Ultimate (Functional) Benefits

Even before the emergence of effective artificial contraception, much of human sexual activity was likely non-reproductive, if contemporary populations are any guide.

Like many other species, men who are sexually frustrated engage in self-gratification. This practice may be simply attributable to the pursuit of pleasure, but it may also improve the viability of sperm in subsequent ejaculates. Masturbation is common among women for whom no functional benefit is clear[3].

In heterosexual relationships, the variety of sexual practices, including oral and anal sex, means that some sexual activity is patently non-reproductive. Of course, such practices may contribute to bonding via shared pleasure and emotional experience.

Even if a couple is restricted to intercourse, most of their interactions occur at times when conception is impossible. This is because women are fully fertile only for a few days at most in their monthly cycle. Then, there is sex during pregnancy and sex after menopause when conception is impossible.

When each of these factors is considered, it is clear that potentially reproductive sex is a small fraction of a person's lifetime sexual activity. What does all this mean?

Conclusion

Even from the restricted perspective of evolutionary biology, sexuality is complex for other species as for humans. Through the ages,

Christian theologians argued that sexuality should only be about reproduction. These ideas now seem absurdly out of touch with reality.

In fairness, the varied functions of sexuality were properly appreciated only in recent decades. They have yet to be placed in a sharp evolutionary perspective.

Sources

1. *Hill, K., and Hurtado, M. (1996). Ache life history. New York: Aldine de Gruyter.*
2. *Starkweather, K. E., and Hames, R. (2012). A survey of non-classical polyandry. Human Nature, 23, 149-172.*
3. *McNair, B. (2013). Porno? Chic! New York: Routledge.*
4. *Roughgarden, J. (2006). Evolution's rainbow. Oakland, CA: University of California Press.*

High Self-Esteem: Good or Bad?

A dark cloud of anxiety and fragility surrounds sky-high self-esteem

A mericans have high self-esteem. Young people, in particular, feel good about themselves. Is this beneficial for health and well-being? Or are we cruising for a bruising?

Participating more in virtual networks, young people are increasingly concerned about how others evaluate them[1]. This is a major source of competition and anxiety that threatens happiness.

Raising Confident Children

Although participation in electronic social networks (which are recent) may boost self-esteem, narcissism has been rising for a long time - at least three decades - so it is hard to claim that self-promotion on Facebook is the only source of high self-esteem amongst college students[1].

American parents strive to raise confident children and do so by avoiding criticism, or any negative evaluation that might dent the ego. Their efforts are bolstered by what happens in schools. The emphasis is very much on schools yielding fun experiences where everyone is reinforced for their efforts, regardless of actual accomplishment or effort. In this Lewis-Carroll environment, all children are winners and they all receive prizes.

Most psychologists would agree that it is better to reinforce children for going to school than it is to traumatize them after the fashion of the teachers depicted in a Dickens novel. Even so, excessive doling out of praise that is not earned by real effort or achievement cannot produce entirely good outcomes.

Ego as Motivation, and as Blindness

One consequence of unearned praise and undeserved rewards is the mediocrity of our educational system. When children are not challenged, they cannot reach their potential. As a result, huge numbers of college students require remedial education to scrape through introductory-level courses.

One beneficial effect of overwhelmingly positive early experiences is that young adults are more willing to take risks, such as those inherent in starting a business or writing a novel. People who feel good about themselves, generally have an optimistic view of their own prospects.

While egoism can be motivating, in general, it can also lead to striking errors of judgment. Such blindness is illustrated by the disastrous Bay of Pigs invasion in Cuba. Planners went ahead with this covert action against the Castro regime as though they assumed nothing could possibly go wrong. Of course, just about everything that could go wrong did go wrong. The invasion site was poorly chosen and strongly defended by the Cuban military, and the popular uprising expected in support of the invasion did not materialize.

These incautious errors brought on the Cuban missile crisis, taking the world close to thermonuclear Armageddon. Fortunately, most mistakes of blind egoism do not have nearly such serious consequences.

Yet, feeling excessively good about oneself can have bad practical results. These range from unwise purchases at the height of an investment mania to getting exploited by fair-weather friends who never pick up the tab at a restaurant. Whatever the costs, narcissism is definitely on the rise. We can point to a variety of reasons[1].

Narcissism Unleashed

Over the past half-century, the lessons from psychology pervaded child-rearing practices. Most obviously, there has been a decline in the use of corporal punishment thanks to more child-friendly policies in opposition to the authoritarian households of the past.

The demise of physical punishment in child-rearing practices represents a more child-centered approach that makes for a happier childhood. It is part of a generally more lenient approach to discipline issues.

One reason that parent-child interactions are more positive today is that families are much smaller and births more widely spaced. This means parents have an opportunity to get to know their kids better and to develop warmer relationships with an emphasis on the use of reasoning rather than punishment. As a result, parents often behave more like servants for their children than as frightening authority figures.

This status reversal contributes to children's sense of power and can boost their self-importance. Arguably, narcissism had a good head start before the Internet emerged.

From the Christmas Letter to the Facebook Page

Narcissists may have a high opinion of themselves, but their self-esteem is often fragile. It is threatened by the possibility of being disliked or rejected by others. Staying ahead of that possibility causes a great deal of anxiety. It motivates much boastful talk that may have a shaky foundation in reality.

Then there is the great effort required by endless self-promotion. We have all groaned at the family-level advertisement that form the substance of many Christmas letters. The genre has now morphed into daily Facebook updates. Pictures constantly uploaded to social media sites brag about lifestyle.

Ironically, those who come across as doing best in their virtual self-presentation are often the most insecure and suffer from the nagging suspicion that other people lead more glamorous and interesting lives than they do.

Source

1. Twenge, J. M., Kenrath, S., Foster, J. D., Campbell, W. K. and Bushman, B. J. (2008). Further evidence of an increase in narcissism among college students. Journal of Personality, 76, 919-928. http://onlinelibrary.wiley.com/ doi/10.1111/j.1467-6494.2008.00509.x/abs...

The Tribal Problem
Are we stuck with tribal conflicts?

J ust as a picture gets divided into figure and background, we divide the social world into us and them. How does that happen? Did it help our ancestors to thrive? Could we stop doing it in the name of peace and harmony?

Evolutionary Views

One view is that tribalism favors group solidarity in times of war. This idea is supported by tribal initiation of warriors.

Initiations that are more costly and painful generate stronger tribal affiliations[1]. The strange bonding effect of suffering is quite general and applies to shared military experiences, painful childbirth, and embarrassing discussion groups.

What is the evolutionary rationale for such bonding? Perhaps groups who experience very hard times need extra bonding to keep them from splitting up. Tribal bonding likely helps warlike societies to persist through difficult times favoring survival and reproduction.

Evolutionary psychologists often assume that making strong in-group out-group distinctions is somehow encoded in our genotype as a pan-human adaptation, but this view has flaws.

To begin with, developmental geneticists find such Darwinian programs biologically improbable and there is no evidence from neuroscientists that they are real[2].

In the particular case of military solidarity, human societies before agriculture rarely, or never, practiced warfare. What is more, their tribal affiliations were weak. Hunter-gatherers had a fluid social structure where individuals could easily leave one group and join another. This practice reduces tensions within groups because the malcontents can leave.

161

Instead of being a fixed element of human societies, strong tribal passions were a functional response to warlike environments. This general principle holds up for modern ethnic groups and nation states.

Nationalist Incitement

Social psychologists have long known that it is very easy to whip up group conflict, but trickier to calm it down. Incitement can be as trivial as giving randomly selected groups a different hat or badge.

Is such conflict built into the human brain? Or, are people responding to characteristics of the situation? When groups are dressed differently, this generally provides reliable information about some material difference. Perhaps they support different sports teams, or belong to different religions, or come from different countries. There are few real-world scenarios in which we meet randomly selected groups who are dressed differently. Far from being built into the brain, such phenomena are acquired by social learning.

When groups perceive themselves as different from others, this often sets the stage for conflict and competition. This after all is a key reason why sports teams wear different uniforms.

Ethnic conflicts fall into the same mold. An Orange Order parade in Northern Ireland looks and sounds very different from a Saint Patrick's day parade.

Apart from such differences of ethnic and religious identity, Catholics and Protestants lived in segregated neighborhoods with the Catholics being economically disadvantaged thanks to government discrimination. The relative peace of the province today rests partly on a deliberate program of neighborhood desegregation.

A path of integration is essential in countries like the U.S. where there are large and varied immigrant populations.

The Melting Pot of Immigration

The study of immigrant populations in the U.S. offers insight into how ethnic gaps get bridged over time. One example concerns the Italian residents of Roseto, a small town in Pennsylvania, who migrated together from their native village[3].

This community came to the attention of health researchers due to its health advantage over surrounding towns. This took the form of better cardiovascular health. Roseto residents had the heart health of average Americans a decade younger or more.

What produced this seemingly miraculous delay of the aging process? Rosetans maintained the social life of their Italian forebears. They spoke Italian and maintained the ceremonies and traditional religious festivals of their Italian ancestors. Health researchers concluded that the magic ingredient of Rosetan lifestyle was social support derived from the community with residents dropping in informally to converse in their neighbors' kitchens.

Although this immigrant community had minimal connections with surrounding communities, apart from those of work and business, matters changed greatly over time as subsequent generations became increasingly assimilated following a pattern found in all other immigrant communities.

This process occurs as children of immigrants get drawn into the lifestyle choices of their peer group in the surrounding community. In Roseto, this involved abandoning the intimate, but cramped, houses of the village in favor of large suburban homes, neglect of traditional celebrations, changes in diet, and complete Americanization of speech, behavior, and interests. It also involved a deterioration in health as cardiovascular problems and related disorders reverted to the national mean.

So tribal barriers are not permanent. They get broken down as the social environment changes. American history shows that assimilation can be painful and typically requires several generations to play out. Of course, that process continues today.

Source

1. *Cialdini, R. (1988). Influence: Science and practice (2nd Ed.). Glenview, IL: Scott Foresman.*

Are Dogs Self-Aware?

Tests may be biased in favor of visual creatures like ourselves

The standard test of self-awareness is being able to recognize ourselves in a mirror. Although chimpanzees pass this test with flying colors, gorillas have inconsistent results. Dogs flunk by treating the reflection as another animal.

Experimentally Self-Aware Animals

Gordon Gallup[1] devised the first credible test for self-awareness. He exposed chimps to a large mirror so that they could get familiar with their own image. A dye mark was surreptitiously placed on the brows of mirror-exposed chimpanzees. The chimps behaved very much as humans might under similar circumstances. They used the mirror to inspect the mark, touched it with a finger, and attempted to remove it.

Few animals pass the mirror test for self-awareness (appropriately modified for species differences in anatomy). They include chimpanzees, bonobos (pygmy chimpanzees), orangutans, at least one elephant[2], dolphins, humpback whales, and magpies[3]. Apart from the magpies, all of these are large-brained animals. They are all highly social as well, with the exception of orangutans, which are mostly solitary as adults.

Magpies are the big surprise on this list, but they and their relatives (the corvids, or crows) are intelligent and pass problem-solving tests that only great apes can master. Results are mixed for gorillas and capuchin monkeys, with some studies reporting that they pass the mirror test, but others reporting that they fail.

Surprisingly, dogs do not pass the self-awareness test. Dogs are highly intelligent, extremely social, and fit right in with human households, even to the extent of voluntarily learning to recognize the meaning of human words.

Anyone who saw the *60 Minutes* segment on border collies knows that these clever dogs are extremely attentive to the needs of their masters. One collie had a large collection of about a thousand stuffed toys that he could retrieve on demand. "Fetch Kermit" always yielded the frog from *Sesame Street*, for instance, and whichever toy was requested, the dog retrieved it. It is hard to imagine that this is anything but intelligent behavior (as opposed to operant conditioning). If so, it suggests that the dog has a clear grasp of the owner's intentions, hinting that a capacity for self-awareness is not unthinkable.

Why Gorillas and Dogs Fail

Inconsistent results for gorillas in mirror self-recognition are sometimes attributed to their relatively small brain size - relative to chimpanzees. Yet, this is a shaky argument. Gorillas demonstrate interest in painting[4], and male gorillas sometimes care for orphans - something that is unknown in chimpanzees[5]. This behavior could be motivated by empathy that suggests self-awareness, although other interpretations are possible.

Gorillas may do poorly at the mirror test because they avoid looking directly at strangers, as this constitutes a threat display. So, it is hard for them to learn that the mirror reflection is themselves.

Dog lovers complain that the mirror test favors visual animals like primates but makes it difficult for dogs, which are more focused on auditory and olfactory cues.

Correlates of Self-Awareness in Dogs

In addition to their general intelligence, as reflected in the many useful tasks that dogs perform for humans (rescuing skiers, herding sheep, silently pointing to prey animals), dogs are socially astute. An ostensibly well-behaved animal might grab a piece of meat from a countertop as soon as its owner's back is turned. If caught in the act, the dog cringes in a way that suggests guilt, or at least fear of punishment. It is difficult to understand these actions without assuming the animal has some sort of mental representation of how it is expected to behave.

Animal cognition researcher Marc Bekoff[6] found his dog Jethro (a neutered Rottweiler mix) could recognize his own scent marks from urine in snow and avoided marking over them, but that is not exactly self-awareness. Indeed, it is likely that all scent-marking animals avoid marking over their own scent in a reflexive way. Better controlled tests replicated Bekoff's result that dogs spend less time sniffing their own scent.

Much as we might wish to believe that man's best friend is self-aware, there is no good supportive evidence as yet, although this may reflect problems with the tests. At this point, all we can claim is that domestic dogs are almost incredibly well attuned to the niche of serving humans. They may have accomplished this so well that we are fooled into thinking they have much the same inner life as we do.

Sources

1. Gallup, G. E. (1970). *Self-awareness in the chimpanzee. Science, 167*, 86-87.
2. Plotnik, J. M., de Waal, F. M. B., and Reise, D. (2006). *Self-recognition in an Asian elephant. Proceedings of the National Academy of Sciences, 103*, 45-52.
3. Prior, H., Schwarz, A., and Gunturkun, O. (2008). *Mirror-induced behavior in the magpie (Pica pica). PLOS Biology 6* (8) e202. doi:10.1371/journals.pbio.006 0202
4. Morris, D. (1962). *The biology of art.* New York: Knopf.
5. Buchanan, G. (2015). *The gorilla family and me. BBC Film Documentary.* http://www.bbc.co.uk/programmes/b06ts2dk
6. Bekoff, M. (2001). *Observations of scent-marking and discriminating self from others by a domestic dog (Canis familiaris): tales of displaced yellow snow Behavioural Processes, 55*(2), 75-79 DOI:10.1016/S0376-6357(01)00142-5

Nigel Barber

The God Spot Revisited
Spirituality arises as an evolved function of the brain

Media coverage hyped the importance of findings suggesting that spiritual experiences have a brain basis. Even so, an evolved capacity for self-transcendence fits in well with what we know about the evolutionary role and function of religion.

The evolution of religion

Religious beliefs and rituals are found in every society studied by anthropologists. This implies that religious/spiritual experience is a universal characteristic of human beings just as the capacity to see in color is.

Religion could not have evolved and could not have affected the lives of the majority of the world's human inhabitants if it had not helped them to solve the problems of surviving adversity and of successfully raising children who would propagate their supernatural belief systems after they had died[1].

So, it makes sense that the brain might be specialized for religious experiences. Indeed, an evolutionary perspective on religion implies that humans are inherently susceptible to religious views.

This view is bolstered by evidence that spiritual experiences (including religious experiences) have a neural basis. Although there is no single "God spot" in the brain, feelings of self-transcendence are associated with reduced electrical activity in the right parietal lobe, a structure located above the right ear[2].

Self- transcendence, or a sense of the otherworldly, is the opposite of being self-focused, and is a convenient definition of spirituality and/or religious sensibility used by researchers. This perception is generated by many experiences in addition to religion, including brain trauma, drug states, and epileptic seizures.

Spiritual experiences use many different parts of the brain: the God spot is functional rather than anatomical. So what are the likely benefits of having such neural mechanisms for spiritual experiences?

What is the God spot used for?

In the past, I have argued that a primary function of religious beliefs and rituals is as a form of emotion-focused coping with the difficulties of life. It functions rather like the security blanket that a small child employs to soothe itself when distressed.

The security blanket concept of religion has a lot going for it. It explains why people pray during a crisis, and why people living in the most miserable places on earth are universally religious. On the other hand, in societies that experience a good quality of life, religion loses its importance, and atheism breaks out[1]. This is what is happening in the social democracies of the world from Sweden to Japan.

Such "comfortable" modern societies are an anomaly, of course. Prior to the emergence of such uniquely favorable conditions, life was always full of difficulties. That is why religion is a human universal. It is also the reason that our religious sensibilities are served by specialized functions of the brain. These snap us out of the self-absorption otherwise induced by misery and produce self-transcendence, or a feeling of otherworldliness.

This is not exactly a God spot because it is neither localized as a spot, nor peculiar to experiences related to a deity. Yet, it adds a dimension to our understanding of religious experience and explains why even people in secular countries remain deeply spiritual[3].

Sources

1. Barber, N. (2012). *Why atheism will replace religion: The triumph of earthly pleasures over pie in the sky.* E-book, available at: http://www.amazon.com/Atheism-Will-Replace-Religion-ebook/dp/B00886ZSJ6/

2. Johnstone, B., Bodling, A., Cohen, D., Christ, S. E., & Wegrzyn, A. (2012). Right parietal lobe-related "selflessness" as the neuropsychological basis of spiritual transcendence. *International Journal for the Psychology of Religion.* accessed at http://www.tandfonline.com on 5/30 2012.

3. Zuckerman, P. (2008). *Society without God: What the least religious nations can tell us about contentment.* New York: New York University Press.

Faith Healing Shouldn't Work, But It Does
How to explain faith healing

Previously, I have discussed whether antidepressants work mainly via suggestion, or the placebo effect. A placebo resembles faith healing. Yet faith healing is usually considered more a matter of belief in magic and the supernatural rather than confidence in the science of pharmacology.

From a scientific perspective, faith healing is unexplained, incomprehensible, and should not work. Yet it does work. The same is true of drug placebo effects, of course. Scientists recognize that there are placebo effects, but have trouble accounting for them.

If you grew up in a superstitious country, chances are you experienced faith healing. Here are some examples from my own childhood in Ireland:

- Children born after their father's death were understood to have the cure for thrush, a throat infection.

- The seventh son of a seventh son had special powers, such as the ability to cure ringworm.

- A cure for warts was inherited in some families.

Such traditional faith healers generally practiced for free, although strangers might wish to compensate them for their inconvenience with a small gift. Given that these services were genuinely free, and given that faith healers considered it immoral to demand payment for their special gift, they were widely used. What of the results?

The proof of the pudding
One year, my sisters and myself became infected with ringworm - a fungal infection that may be acquired from contact with farm animals.

The man with the cure was a local bachelor farmer who could be encountered early in the morning harvesting mushrooms in our pasture. He welcomed us to his cottage and treated our ringworm by drawing a wedding ring across each lesion, making the sign of the cross. "They should be gone in a month," he said. Sure enough, all disappeared in about three weeks.

A close friend had a similar experience with warts. The faith healer knotted pieces of knitting wool above each wart, without touching it, while reciting a Hail Mary. The warts fell off within a month.

Most scientists cope with such evidence through simple skepticism. Perhaps the ostensible "cure" had no connection with the outcome. Without treatment, the time course of recovery would be exactly the same. It is certainly true that ringworm undergoes spontaneous healing. This is a seasonal phenomenon, however, with the rash characteristically flourishing during wet, or humid, seasons and spontaneous recovery would have required several months, not a few weeks.

The girl had also had her warts for at least two years, so that their accidental recovery in a month was even more unlikely.

It is always hard to make much sense of such anecdotal phenomena to the satisfaction of scientists, but faith healing seems to evoke a placebo effect, not unlike the use of drugs to treat people who are mildly depressed (and therefore experience no true pharmacological response to the medicine).

When people receive a prescription drug, such as Zoloft, or Paxil, they expect improvement making them fair game for a strong placebo response. Why should recipients of faith healing expect to get better? Several elements of the situation conspire to give patients the expectation that they will get better.

To begin with, there is the mumbo jumbo about which individuals acquire the gift to heal a specific malady. Notice how the pagan aspects of faith healing, or "superstition", are combined with Christianity to convey the impression that different supernatural forces are working on the problem. Social pressure to believe in the cure might also be a factor - after the manner of The Emperor's New Clothes.

If there is a history of successful outcomes, then people who consult the faith healer are likely to show up because they already have a positive expectation of cure, even if they consider themselves too sophisticated to be taken in by magical thinking.

By means unknown, faith healing is evidently capable of boosting immune function. This would explain why minor lesions clear up faster than would otherwise be the case. If placebos account for half of the effects of non-surgical medicine (which may be too conservative an estimate), faith healing may be a trillion-dollar industry in the U.S.

Why Waitresses Look Sexy
Men give larger donations to attractive women

There is a basic grammar of courtship in birds where the male offers a gift to the female, such as a bit of food, or a piece of nesting material. If the gift is satisfactory, the pair moves on to the next phase of bonding. This grammar also influences men's willingness to donate to women, even in an era of gender equality.

The Attractive Female Effect on Donations
Giving to charity is a surprisingly emotional phenomenon. In experiments, people who are asked to think rationally before making a donation, by solving a simple math problem, for example, make smaller donations. Also, men give more to female collectors, and they give most if solicited by an attractive woman according to English research[1]. On the other hand, women do not give more to male collectors (handsome or otherwise).

As to why men are more likely to give to women, there is an obvious courtship analog. In virtually all societies, men give gifts to women as a prelude to courtship. This might be an ornament, a piece of clothing, or some food.

Amongst hunter gatherers, a successful hunter might hide some meat when the game is divided up and send it to a potential lover via an intermediary. In developed countries, male suitors may pay for a meal in a fancy restaurant as a way of initiating a romantic relationship. In an age when many women are willing to pick up the tab, such customs retain an emotional significance that is hard to escape. Why might an attractive woman collect more for charity? Why do sexually attractive waitresses get bigger tips from men?

The Sexy Waitress
The logic of male generosity in a courtship situation is complex. It is more than a token offering through which male birds can defuse

aggression in a female by stimulating a different behavior, such as feeding, or nest-building. Other reasons to be generous as a tipper include:

- Wanting to seem kind, sensitive, and considerate.

- Advertising earning capacity and social status by giving freely.

- Placing the waitress under an obligation that could be repaid sexually.

Intercourse is widely considered a privilege that women provide to men. That is not always true. One clear exception is the case of middle-aged British women who visit Kenya as sex tourists. They hook up with much younger local lovers and reward them with gifts. Of course, young attractive women do not need to give gifts to anyone!

Male generosity is also an index of the man's social status. If a man gives the waitress a large tip, he is also advertising his own social status in terms of disposable income. It just happens that women weigh a man's earning capacity more highly as a mate selection criterion than men do.

When a man is served by an attractive waitress, he is rather unlikely to strike up a romantic relationship with her. Even so, her income will improve by dressing sexily. As the English researchers said, "Good deeds among men increase when presented with the opportunity to copulate."

Source:
1.https://www.philanthropy-impact.org/news/men-more-generous-when-attractive-women

What Does GameStop Have to Do With the Insurrection?

The Insurrection arrives at Wall Street

On the surface, the financial attack by a group of investors on Wall Street had little to do with the January 6th physical attack on the Capitol. Yet, both movements involve a fanatical assault on powerful institutions by alienated Internet-organized groups.

GameStop and the Power of the Internet

GameStop's business model relies upon brick-and-mortar sales of electronic games at a time when most of these sales are online. It is perceived as a dinosaur business that will continue to decline after the fashion of Blockbuster and other video stores.

Its demise was being hastened by short-sellers who bet that the stock price would decline, in which case, they would pocket most of the value of that stock decline. Short selling is a risky practice because a stock can rise without limit, but it can only fall to zero.

Even worse, when a shorted position rises precipitously, the owner of the short position is forced to raise more money to cover the loss. That is why the GameStop mania brought at least one hedge fund to its knees, forcing it to be bailed out.

GameStop's main asset propelling its stock price higher was a nostalgic appeal to millennials who recalled happy hours browsing its shelves. Many felt that this icon of their adolescence was being unfairly targeted by Wall Street wolves.

These views were shared on the Reddit forum WallStreetBets and crystallized into a rebellion analogous to the Occupy Wall Street movement of a decade ago. While this earlier revolt against the Wall Street establishment fizzled, the Reddit movement had an advantage in numbers with up to six million people involved.

They also knew that GameStop was heavily shorted and that any concerted buying would spook the shorts, forcing them to buy the stock, thereby pushing it up even further in what's known as a short squeeze.

This short squeeze was all too successful and drove GameStop prices into the stratosphere. It effectively came to an end when popular trading platforms like Robinhood either restricted purchases in the stock or temporarily shut them down. The trading volume in GameStop was so high that the platforms feared running out of funds, or so they claim.

So far, the revolt against Wall Street resembles the insurrection in which an online community targeted a national institution at the height of the COVID crisis, when many had time on their hands. In each case, there was a strong feeling that institutions like Wall Street and Congress serve elites and work against ordinary people.

The Assault on Reality

When people get absorbed in a mass movement, mob psychology takes hold and individuals often behave far more recklessly than they would alone. The first casualty is the perception of reality.

Whether it is a religious cult or an Internet stock-trading group, there is a tendency to reinvent reality, to float truths that are peculiar to that particular group of people.

At the height of the GameStop mania, WallStreetBets participants were bragging about how much they had paid for their shares and vowed to hold on with "diamond hands." Conventional stock traders aim to buy low and sell high, rather than buying high and being ready to lose all.

The Capitol insurrectionists began from the premise that the election had been stolen. From there they proceeded to "defend America" by attacking the seat of democracy at the time it was formalizing the most important ritual of democracy, namely the peaceful transfer of power according to the wishes of the electorate.

One of the more jarring aspects of both insurrections is that many of the participants were far from being the underprivileged people who stoked rebellions throughout history, from ancient Rome to the French Revolution.

The Middle-Class Rebel

In the January 6th Capitol riot, many of the attacking mob were pillars of the community – business owners, professionals, even police.

Likewise, the Reddit stock forum was not an underprivileged group. Some of the WallStreetBets community had hundreds of thousands of dollars at their disposal. Of course, their impact on Wall Street was magnified by the unprecedented size of the disaffected group.

Perhaps the oddest aspect of the GameStop story is that professional traders and hedge funds were attuned to what was happening. Some exploited the mania to their own advantage and have taken to monitoring the Reddit group to identify new stock manias.

This is nothing new. Prior to the dotcom bust in 2000, message boards were manipulated by Wall Street insiders in various pump-and-dump schemes.

Much has been made of both rebellions being "unprecedented" and defining a new era, although one was violent and criminal and the other was peaceful and legal. We do not know what the future will bring but, for the time being, both rebellions shook institutions - they did not topple them. But powerful institutions have been warned: they need to show that they do not just care about the one percent.

The Secret of Creativity: an Oblique Perspective
Why immigrants and gays are so creative

Creative people are complex, meaning that they see the world from multiple perspectives. This is an adaptive response to complex inputs during childhood. We are all constantly trying to make sense of the world we live in and the more complex our experiences, the more challenging this proves to be. This challenge is the key to creativity.

Biographically speaking, creative people have a foot in two camps. In the U.S., for example, immigrants are seven times more likely to excel in creative fields compared to individuals whose families have lived here for generations[1].

Creative people also tend to have a foot in either gender camp. Many highly creative people are androgynous (masculine women or feminine men). Individuals who score high on tests of creativity are more likely to be androgynous as assessed by the Bem sex roles inventory, meaning that they score high both on masculinity and on femininity scales[2]. Androgynous individuals may be heterosexual or homosexual.

Sexual orientation is itself a factor in creativity with homosexuals (male and female) being over represented in most creative endeavors. Cities in the U.S. with a high proportion of gay residents, as inferred from Census data on living arrangements, also have a large proportion of residents working in creative industries such as research, publishing, design, music, film, television, advertising, fashion, theatre, and so forth[3]. Presumably, gay people are attracted by creative opportunities, but it is also possible that the tolerance and openness of highly creative cities is particularly welcoming to homosexuals.

Being an immigrant, being androgynous, or being gay, are all dimensions of otherness, of not quite fitting in with mainstream social categories. Just as people from different countries perceive social interactions differently, so males see the world from a different perspective than women. These gender filters are a complex amalgam of biology and social amplification.

Whereas the strongly gender-typed individual sees the world through a single filter, the androgynous individual can perceive the same event simultaneously through masculine and feminine eyes. An androgynous man is thus analogous to a male immigrant in the territory of women whereas an androgynous woman is like an immigrant in the nation of men. Gays are like immigrants in the world of heterosexuals.

Anyone with such an oblique perspective is privileged when it comes to artistic creativity. An immigrant, or an androgynous individual is more likely to see the same event as having opposite connotations. An ethnic joke that ridicules one's ancestry is simultaneously amusing and painful, for example. Similarly, an androgynous person is quite capable of enjoying a horse race and simultaneously grieving for the abuse meted out to human and equine contestants.

If a person naturally associates opposites in this way, they are very good at dredging up a large number of unusual mental associations which increases artistic productivity and complexity. This is called "divergent thinking." It is what tests of creativity assess by asking people to think of many different uses for a common object such as a brick, for example. Creativity is largely environmental and has only a small genetic component.

A person in the mainstream thinks more simply, and more convergently, i.e., using straightforward logic. They have trouble associating dissimilar images, and the associations they do make are predictable, circumscribed, and conventional. That is why people raised in comfortable homes tend to be intelligent, successful and happy, but to lack creative drive[4]. In Terman's classic study of intellectually gifted children many from affluent homes, for example, not one achieved prominence in any creative field.

A person does not have to be an immigrant, or a homosexual to be creative, of course, but something else gives them that oblique perspective. It might be childhood illness, or the loss of a parent, or some other experience that allows them to perceive the world differently from the mainstream.

Sources

1. Goertzel, V., Goertzel, M. G., & Goertzel, T. G. (2004). Cradles of eminence: Childhoods of more than 700 famous men and women. Scottsdale, AZ: Gifted Psychology Press.
2. Jonsson, P., and Carlsson, I. (2000). Androgyny and creativity. Scandinavian Journal of Psychology, 41, 269-274.
3. Florida, R. (2002). The rise of the creative class. New York: Basic Books.
4. Terman, L. M., and Oden, M. H. (1959). The gifted group at mid-life, thirty-five years follow-up of the superior child. Stanford, CA: Stanford University Press.

What It Means to Be a Human
Humans are surprisingly difficult to pin down as an evolved species

From Darwin onwards, scholars have struggled to define our species. There are two leading theories, but neither seems workable. One sees human psychology as shaped by evolution and stuck in the past. The other defines us as a cultural species that learns how to be human from scratch.

The Evolutionary Psychology Approach

Evolutionary psychologists claim that people across all societies behave similarly and emphasize genetic causes for this similarity. For instance, men are more interested in casual sex, more physically aggressive, and more willing to take risks compared to women. Likewise, jealousy is a leading cause of spousal homicides everywhere, and young adults are perceived as most sexually attractive.

From an observational perspective, such claims ring true. Accounting for them is more problematic. Evolutionary psychologists have asserted that, during our more than two million year history, we became adapted to a hunter-gatherer way of life that favored such behavioral, and psychological, attributes. Genes that predisposed us to such traits were favored by natural selection.

But developmental genetics does not support this: genes cannot encode behavioral or psychological programs[1].

We must also bear in mind that the environment changed dramatically many times over the past two million years, in terms of climate, subsistence economy, and social structure. Moreover, ancestral humans were very different in terms of body size, brain size, anatomy, and thermal physiology, not to mention breeding system and social complexity.

Humans encompassed many different species, including Australopiths (i. e., "ape men" like fossil Lucy), Homo habilis, Homo erectus, Homo sapiens, and the many dead-end species still being discovered. The "environment of evolutionary adaptedness" is a tall tale because each species encountered varied environments.

If evolutionary psychologists focus on supposed adaptations to past conditions, cultural determinists define human beings in terms of information that is received via social learning in the lifetime of the individual.

The All-Cultural (Blank Slate) Approach

To a cultural determinist, we are defined not by our genetic heritage but by what we learn as members of a community. Examples include languages, religions, subsistence practices, and tool manufacture. This approach to defining humanity is also problematic.

If socially learned information defines us, what is information? Is it the verbal response to some item on a questionnaire, such as, "I approve of premarital sex"? Or does it consist of changes in our brain cells that take place when such information is acquired? Or, is it a packet of information that is capable of being replicated? Theorists have not been able to nail down a definition that inspires widespread consensus.

Factually speaking, social learning is not peculiar to humans. Indeed, social learning is probably a feature of all social vertebrates[2].

Perhaps for this reason, anthropologists insist upon the cumulative quality of human social learning as our defining feature.

The argument is that human societies contain much more information than can be mastered by any one individual[3]. This point is underlined by the information age where the volume of data is increasing at a phenomenal rate, but it has not always been true.

In simpler societies, individuals can actually master most of the technological expertise, and other transmissible knowledge in their society. Moreover, historical study of artifacts, such as arrowheads, finds that their designs are transmitted across generations, rather than being spread horizontally across populations (or "diffused"[4]). This means that cumulation of social knowledge is quite recent, most likely arising after the Agricultural Revolution.

So, if we want to define humans as unique in having cumulative socially learned information, we would have to exclude most hunter gatherers from the category of humanity.

The main scientific problem with the cultural species approach is that it sets humans apart from the natural world by claiming that we are shaped by a specific society rather than by natural selection. Evolutionary psychologists argue that although mismatched to our current environments, we are adapted to a (fictitiously uniform) ancestral one.

Presented with two bad alternatives for defining our place in the natural world, it is reasonable to look for a better alternative, namely that humans, like all other species, are adapted to their current environment.

A Species Well Adapted to Modern Conditions?

There are many different ways in which animals become adapted to their current environments, but evolutionists over-emphasize genetics because this fits most easily with Darwinian theory. Research suggests that genetic determinism has little to do with complex behavior although there are certainly genetic effects on temperament and personality.

Even simple adaptive behaviors are not genetically transmitted. Moose do not come into the world being afraid of wolves, their natural predator - they have to learn this, from their mothers and from experiences[4].

Even if one restricts the focus to genes, humans (and other species) can become adapted to varied local conditions with remarkable rapidity. One intriguing example involves alcohol intolerance in people from rice-growing regions of Asia. With rice farming, it was all too easy to make large quantities of alcohol. This posed a significant risk of alcoholism which was better resisted by individuals who were alcohol intolerant, and therefore left more offspring[5]. Another example is the evolution of lactose tolerant adults in places where dairy farming was widely practiced. Such changes occurred within the last 5,000 years.

The pace of human adaptation has been even more rapid since the Industrial Revolution. Some obvious differences include increased stature (by up to 20%), rising IQ scores (about 30 points in developed countries), increased life expectancy at birth (by about 100%), increased standard of living in terms of work hours needed for subsistence, and the decline of both marriage and fertility - to about a third of agricultural levels[6]. Such changes are not due to gene selection, of course. Nor are they all beneficial (increases in allergic diseases, obesity, diabetes, etc.). Yet, all are a reaction to environmental change.

What it means to be human is very much a moving target. Both leading narratives are so far out of touch with reality that they look like badly scripted science fiction movies. We need to investigate adaptation to contemporary conditions - and go beyond cultural, or genetic, determinism which explain little about what it means to be human.

Sources
1. Carroll. S. B. (2005). *Endless forms most beautiful: The new science of evo devo and the making of the animal kingdom*. New York: W. W. Norton.
2. Richerson, P. J., and Boyd, R. (2004). *Not by genes alone: How culture transformed human evolution*. Chicago: University of Chicago Press.

3. Mesoudi, A. (2011). *Cultural evolution: How Darwinian theory can explain human culture and synthesize the social sciences.* Chicago: University of Chicago Press.

4. Berger, J., Swenson, J. E., & Persson, I. L. (2001). Recolonizing carnivores and naïve prey: Conservation lessons from Pleistocene extinctions. *Science, 291,* 1036-1039.

5. Henrich, J. (2015). *The secret of our success: How culture is driving human evolution domesticating our species and making us smarter.* Princeton, NJ: Princeton university Press.

6. Floud, R., Fogel, R. W., Harris, B., & Hong, S. C. (2011). *The changing body: Health, nutrition, and human development in the Western world since 1700.* Cambridge, England: NBER/Cambridge University Press.

Anxiety Springs Eternal

Anxiety is key to basic survival mechanisms but is triggered by social life

A worm lives in the apple of human happiness. That worm is anxiety. However accomplished, happy, and admired a person might be, the well of anxiety never runs dry. It is the unsolvable problem that gives rise to many others, from addiction and suicide to depression, obesity, and economic failure.

To Live is to be Anxious

Wild rabbits nibble nervously with constant alertness to movements in the background, to sounds, and to the odor of predators. Their anxiety is valuable because - as a popular prey species - it keeps them alive.

Subsistence humans had fewer predators, but they had to be on constant alert for large cats, poisonous snakes, or biting insects that were a serious threat to adults and children alike.

Of course, humans were top predators in addition to being prey. Even top predators like lions have something to fear, and it is not uncommon for lions to be attacked and killed by a herd of water buffalo defending their young.

There is some safety in numbers and humans. Prior to the Agricultural Revolution, humans lived in groups of over forty individuals. That helps explains why our species is so highly social. When we communicate our fears to others, we typically feel that our worry is lessened.

Social living minimizes both real dangers in the environment and also imagined ones. Ironically, social interactions are themselves a major source of anxiety because they can affect the social support available for difficult times.

The Social Dimension of Anxiety

Social interactions mostly contribute to feelings of calm and security. This mood is strengthened by close social contact that stimulates the release of oxytocin (the cuddling hormone) in much the same way as interactions between mothers and offspring[1]. So, contact with trusted friends and associates reduces anxiety.

In developed countries, families are smaller which means we may spend much of our time surrounded by strangers. This is potentially anxiety-provoking.

Franz Kafka described an alienating world where the individual is dwarfed by a totalitarian state that remains distant and unknowable. He was indirectly describing modern life in which the local community has evaporated, leaving us vulnerable to distant and alien authority.

Whatever one thinks of the paranoid fantasy world of a Kafka novel, he touched a chord by depicting a political landscape where the individual is isolated and powerless.

Although social interaction may relieve anxiety, interacting with large numbers of people may also be a source of anxiety, particularly if many of them are strangers.

Perhaps that is why social events are so often accompanied by alcohol consumption. Of course, in our society, alcohol is the most popular anti-anxiety drug. Other drugs in the same category exert similar effects on the brain - they suppress activity in the cerebral cortex thereby relieving feelings of anxiety and inhibition.

Given the important role anxiety plays in alerting us to the dangers around us - whether physical, or social - it follows that the most popular drugs of abuse are anti-anxiety drugs and medicines (including sedatives, muscle relaxants, and sleeping pills).

Opiates like heroin and oxycontin are in a different drug category, but feature powerful calming effects. Marijuana is also used to help people chill out even though it is classified as a hallucinogen.

Suffice it to say, if anxiety were not such a prevalent experience, large segments of the population would not be addicted to drugs, whether prescription or recreational.

Health Costs of Chronic Anxiety

Drug addiction is a huge and growing problem. Untreated anxiety is a threat to health and longevity as well. Some sense of the scale of these problems is derived by looking at the range and scope of its effects on mental disorders and health.

Although treatable by non-chemical interventions, anxiety disorders are the most common ones in the sense that most people experience fears that are out of proportion to the actual threat posed. They may fear flying, or dread the number 13. Severe obsessions and compulsions are much less common.

Anxiety may be a problem in itself, but it is even more significant as a contributor to more serious illnesses.

Chronic anxiety contributes to clinical depression which in turn may contribute to heart disease in the sense that diseases are biochemically related[2]. So, if the problem of anxiety disappeared tomorrow, a lot of clinical psychologists and psychiatrists would be out of work.

The same is true of many other medical specialists. Beginning with heart disease (still the leading killer in many developed countries), the toll includes common gastrointestinal diseases that are aggravated by stress, and reaches out to the entire metabolic syndrome (i.e., obesity-related disorders) which are the single most important health threats in developed countries.

189

Unhealthy eating patterns can be a reaction to anxiety that exacerbates the risk of obesity in sedentary population, particularly those who are disadvantaged by social inequality[3].

While social occasions are often accompanied by anti-anxiety drugs, chemical solutions to anxiety are not the only ones. They mostly yield further health problems. Because anxiety is built into us as a mechanism of self-preservation, it is going nowhere.

Sources

1. Uvnas-Moberg, K. (1998). Oxytocin may mediate the benefits of positive social interaction and emotions. Psychoneuroendocrinology 23: 819-835.
2. Dhar, A. K., and Barton, D. A. (2016). Depression and the link with cardiovascular disease. Frontiers in Psychiatry, 7, 33.
https://www.ncbi.nlm.nih.gov/pmc/articles/PMC4800172/
3. Wilkinson, R., & Pickett, K. (2010). The spirit level: Why greater equality makes societies stronger. New York: Bloomsbury Press.

Lightning Source UK Ltd.
Milton Keynes UK
UKHW020752020821
388172UK00012B/1012